Letting God Create Your Day

Volume 2

Paul A. Bartz

CREATION MOMENTS

Letting God Create Your Day, Volume 2
by Paul A. Bartz

Fourth Edition, 2009

Creation Moments, Inc.
P.O. Box 839 Foley, Minnesota 56329
www.creationmoments.com
800-422-4253

ISBN 1-882510-18--6

Cover Photo of the Trifid Nebula
Image credit to NASA/JPL-Caltech/NOAO

Printed in the United States of America

Foreword

Few would question the fact that we human beings seem to come into this world having three eternal questions burned into our souls: Where did we come from? Why are we here? And where are we going? All of us, at some time during our lives, ask one if not all three questions. Some dismiss them rather quickly, others linger over these questions for years. Nevertheless, those questions persist, and not just in this generation. They have occupied the minds of thinking people throughout recorded human history.

At the very beginning of history, in what we now call the Old Testament, may be found the most complete account of the beginnings. These writings were preserved by a handful of Jewish priests; they answer the first two questions and hint at the third. They ascribed the origin of the entire universe and all life - including human life - to an intelligent Creator God. The coming of Jesus Christ gave the world the New Testament, which not only clarified those first two answers but also presented the solution to the third question. Humanity's origin, purpose and destiny could now be known with certainty.

Twenty-three centuries ago the Greek philosophers gave us their answers to these three eternal questions, and the Greeks ascribed the origin of everything to time and chance, effectively denying the Creator God. Time and chance leave no place for purpose or destiny. The Greek explanation lay dormant until the last century when, with impetus from Charles Darwin, it was popularized and came to be known as evolution. Today this anti-Christian notion is popular but is actually being challenged by the discoveries of science as it has never been challenged before.

When the Apostle Paul went to Athens to discuss matters of human purpose and destiny with the Greek intellectuals, he began with origins. The account may be found in Acts chapter 17. Paul knew this approach was necessary, because without a proper understanding of where we came from, there would be little hope of understanding either purpose or destiny. Today, the theory of evolution has so gripped the human mind that we face a situation identical to that which Paul faced in Athens.

In 1987 another Paul - Paul Bartz - began the *Creation Moments*™ daily radio program. The object then, as now, was to provide evidences from science for the work of the Creator. Once recognized, the existence of a Creator provides purpose for living and hope for the future. I suggest that this very purpose and hope, denied to the average person for so long, is the secret of the success of that little two-minute program. As the voice of *Creation Moments* since 1996, it is my privilege to continue where Paul Bartz left off.

Creation Moments is now broadcast daily in English on more than 1200 radio outlets throughout the world and is also sent out in a daily email. In addition, it is regularly translated and broadcast in numerous other languages, including Spanish. Literally thousands of people have responded by telephone or letter with wonderful testimonies of having been built up in their faith in the Creator, or having actually been brought to a saving knowledge of the Creator. Making copies of each day' s script available to those who call in was a service that enabled callers to share the message with friends. However, it became evident that listeners were asking for something more, and so the *Letting God Create Your Day* series was produced.

Ian T. Taylor
Host of *Creation Moments*

Preface

Thousands of *Creation Moments* programs have now been broadcast around the world. Many thousands of copies of books like this one comprised of *Creation Moments* scripts are in use. We at Creation Moments, Inc., are thankful to every one of you who have written or phoned in your encouraging comments about *Creation Moments*.

Many *Creation Moments* listeners have told us how they have used what they learned on the program or in other books. Many use them as family or Sunday school devotions. Home schoolers have used books of *Creation Moments* scripts to help make science more interesting. Some have told us that they have shared what they have learned from Creation Moments, Inc., with both Christian and non-Christian friends.

All of these examples illustrate that science is not a dry, distant subject for the Christian. Science is fun and interesting. It surrounds us. Science begins to happen every time *you* are curious about something you see, hear, or experience. We at *Creation Moments* are pleased that so many young people have become excited about science through *Creation Moments* radio and the publication of these scripts.

I personally believe that many young people have been turned off to science because of evolution. Evolutionary claims have led many of them to believe that science is the study of things that defy logic and happened "millions of years ago." Who would want to spend a lifetime studying events that happened in the distant past, especially when those events make no sense in terms of modern experience?

Nor is science meant to be a subject we study so that we can show atheists that the Bible is true. I'm glad to see that I'm not alone in experiencing a bit of exhilaration when I encounter some of the more astonishing designs in the creation. I, too, often exclaim, "Now how can anyone reject the Creator in favor of evolution in view of this!" However, where there is unbelief there is a will; where there is a will there is a way. We can never forget that neither science nor logic will ever bring someone to the saving knowledge of Jesus Christ.

However, knowledge of our Creator's design and power in the creation can cause unbelievers to question their "comfortable," Creatorless world. The Holy Spirit can use this questioning, along with a powerful witness to the truth, to cause a person to be dissatisfied with a Creatorless, and therefore impersonal, universe. At this point the Word of God can be heard offering its meaningful answer to the person's needs. The Holy Spirit can use this situation to work saving faith in Jesus Christ.

Neither is science a field of knowledge to fear. No one who is fully aware of the evidence can believe in evolution instead of the Creator solely on the evidence. Belief in evolution is, most simply, a religious belief chosen because one wishes to avoid a personal Creator. This means that no scientific discovery will knock the foundation out from under the truth of God's Word! We need to keep this in mind every time some scientific "authority" claims to have proven that the Bible's view of the universe is obsolete.

Rather than fear science, true knowledge of the wonderful workings of our loving Creator leads us to be overwhelmed by a desire to praise Him!

May you be led to praise your Creator for His wonderful work as you read these pages. May you be led to trust the saving work of Christ for you as you see in these pages His immeasurable love for *all* that He has made!

Paul A. Bartz

The Not-So-Blind Watchmaker

Luke 6:39

"And He spake a parable unto them, Can the blind lead the blind? Shall they not both fall into the ditch?"

Several years ago a book was published that tried to answer the "watchmaker" argument for a Creator. Simply put, the argument says that if you find a watch, you know that the watch didn't just happen. There had to be a watchmaker. This book, entitled *The Blind Watchmaker*, tried to show that the design we see in the creation is simply an illusion—there is no Watchmaker.

In answer to this claim we can point to the millions of wonderfully intricate designs in the creation. However, it seems more fitting to respond to the claim that there is no Watchmaker by mentioning the many "timepieces" in the creation that tie everything together.

All of us are aware of the daily rhythm of waking and sleeping. However, this is only one of many timepieces in creation. Our bodies also have weekly, monthly and annual rhythms. So important are these rhythms to life and health that physicians are beginning to learn how to use them for healing. Doctors know, for example, that jogging in the evening will reduce platelet clumping in the blood; however, it won't in the morning. And doctors have learned that T-cells, which fight infections, are at a peak in the wintertime. Ulcer medicines are now timed to release in the body at night when flare-ups are the worst.

When medicine is so reliant on the many clocks built into us, we know they are not illusions. The entire complex creation is tied together by an even more complex set of clocks—all telling us that the Watchmaker is *no* illusion!

Prayer: Dear Father in heaven, I thank You that in Your mercy, You allow us to learn about how You made the creation operate so that we can alleviate suffering and death. I pray that this knowledge would lead people to You and not away from You. In Jesus' Name. Amen.

REF: Rosenthal, Elisabeth. 1990. "Medicine a matter of timing," Minneapolis *Star-Tribune*, October 2.

Devious Fungus

Job 5:8-10

"I would seek unto God, and unto God would I commit my cause: Which doeth great things and unsearchable; marvelous things without number; Who giveth rain upon the earth, and sendeth waters upon the fields:"

Parasites, especially fungi, are usually written off by evolutionary biologists as simple, primitive organisms. However, whether they choose to recognize the Creator or not, biologists are learning that nothing in God's creation is simple or primitive. Parasites seem to display some of the most clever planning or scheming—depending on your point of view—in all creation.

A parasitic fungus that infects a roadside weed called catchfly *seems* to do nothing harmful to the plant. After infection, a profusion of white blossoms will open—even a few weeks earlier than normal. Bees and butterflies will arrive to collect the grains from the flowers. However, the grains they are spreading aren't pollen but fungus spores. The fungus has not only turned female flowers into males, but has transformed the stamens into spore factories.

This is not the only trick played by the fungus. It also coaxes the catchfly into producing far more flowers than usual, making the plants it infects more attractive than surrounding catchflies. As a result, pollinators come first to the infected flowers before going to the uninfected plants.

This is but one example among many showing that nothing from God's hand is carelessly done. The lesson for us is more than that living things don't fit into the evolutionary simple-to-complex scheme. God also wants us to know that He cares about everything He has made, no matter how lowly.

Prayer: Dear Father, if You have so provided for the lowliest in Your creation, I know that You have provided much more richly for me. Let me be reminded of this especially when I am tempted to worry and act as if I were in charge of taking care of myself. In Jesus' Name. Amen.

REF: Cowen, Ron. 1990. Parasite power. *Science News*, v. 138, October 6. p. 200.

Fabulous Ancient Science

1 Corinthians 15:49

"And as we have borne the image of the earthy, we shall also bear the image of the heavenly."

For centuries, the Chinese were the world's greatest scientists and engineers. Many people know that the Chinese invented the compass and gunpowder, but these are only the beginning. Over 1,000 years ago, Chinese astronomer Su Song built the first planetarium. He even tilted it at a precise 23.5° to match the earth's tilt. More than 2,000 years ago, another Chinese astronomer noted the moons of Jupiter—long before Galileo "discovered" them.

Almost 1,000 years ago, the Chinese used gunpowder to make over 30 types of fire arrows, including one that was a two-stage rocket. They were great shipbuilders as well. One of the most famous examples is a nine-masted, 300-foot ship with watertight bulkheads. The Chinese-invented odometer featured reduction gears and carved figures that beat a drum every time a distance of about one-fifth of a mile was traveled.

The world's first known working seismometer—a device that registers earthquakes—was built in 132 AD. A 16th-century palace roof was designed with a series of interlocking joints that were layered to provide flexibility in case of an earthquake.

As our knowledge of humanity reaches into history and grows, the picture we are seeing is not that of a primitive becoming more like a modern human. Rather, we see humans—ever curious and inventive—creatures that are very different from the animal world. Truly we have come from the hand of God Himself!

Prayer: Lord, I have yet to begin realizing the wonderful abilities You have given me. For this, forgive me. Teach me not to seek for myself what You already provide, but rather to develop those things within myself that You have given me to develop in service to You. Amen.

Bullet-Proof Spiders

Colossians 1:15-16

"Who is the image of the invisible God, the firstborn of every creature: For by him were all things created, that are in heaven, and that are in earth, visible and invisible, whether they be thrones, or dominions, or principalities, or powers: all things were created by him, and for him:"

The silk produced by spiders is very different from the silk produced by silkworms in some very important ways. While silkworm silk is strong yet not prone to stretching, spiders make silk that is both strong and stretchy. They also make several different kinds of silk, each with its own purpose.

In fact, spider silk is stronger than an equal amount of steel and yet it stretches. Spider silk is made of nothing more than proteins. However, scientists are finding it extremely difficult to study the chemical structure of spider silk because it resists most of their efforts to break it down into its chemical parts. Scientists have finally resorted to hot, concentrated acids to break down the silk. They are learning that one type of spider silk is made up of lengths of very stiff amino acids, separated by segments of amino acids that form a springy helix.

Together they contribute to the strength and stretchiness of spider silk. Scientists are exploring how spider silk is made, because this wonderful material promises new methods for creating stronger and lighter materials for human use. One possible use would be making cloth that is both comfortable and bulletproof.

God has done all things well. The fact that we can learn to design better things by learning how He designed things is a witness for His wisdom as well as a witness against claims that the creation designed itself by trial and error. However, there is much more to God's plans for us, as you can learn from the Bible.

Prayer: Father, I praise You, for you have wisely and wonderfully made all things. I especially praise You for Your love and plan of salvation for me through the forgiveness of my sins because of what Jesus Christ has done for me. In His Name. Amen.

REF: Amato, I. 1990. "Unraveling the biochemistry of spider silk." *Science News*, v. 138, Oct. 6. p. 214.

Bumblebee Flight Still a Mystery

Psalm 100:3

"Know ye that the LORD he is God: it is he that hath made us, and not we ourselves; we are his people, and the sheep of his pasture."

What keeps the big-bodied bumblebee with those little stubby wings in the air? Scientists are still trying to learn the secret of bumblebee flight. And while they are learning some astonishing things about the bumblebee, they still don't know how he does it.

Scientists have been testing bumblebees by monitoring their energy use, placing them in wind tunnels, and measuring their oxygen use. No, they haven't been able to make little oxygen masks for the bees—bumblebees have 24 breathing holes.

Scientists have learned that a flying bumblebee flaps its wings 160 times per second and consumes the equivalent of 180 candy bars per hour! A flying bee uses the same amount of oxygen for its body weight as a flying bird or a bat. However, unlike birds, the bumblebee uses no more oxygen when hovering than when flying. This is only one of the discoveries that contradict scientists' theories. In fact, the director of the latest studies on bumblebee flight admits that they still have no idea how the bumblebee stays in the air. He warns other scientists to stop using the current theories on the subject because they are far too simple.

Of course, our Creator God knows how to make a bumblebee fly, even if the best of modern science can't figure it out. This should help us to see that when scientists tell us they have figured out how everything could have been made without a Creator, they don't know what they're talking about.

Prayer: Dear Lord, I thank You for the wonders and mysteries in the created world, for they teach us to remember our place as creatures in Your creation. I also thank You that I am a redeemed creature of Yours. Amen.

REF: Cowen, R. 1990. "Bumblebee energy: What's the buzz?" *Science News*, v. 138, Oct. 6. p. 215.

The Tropical Arctic

Psalm 119:89

"For ever, O LORD, thy word is settled in heaven."

Both archaeology and the Bible suggest that the world was once quite different from the world we know today. Each agrees that the earth's climate was once much more uniform than it is today.

At one time the tropics may have extended into what are today polar regions. However, the Bible and evolutionary geology don't agree on how long ago that was. Bible-believing geologists, on the other hand, are reporting evidence to support the Bible's story of history.

In the last few years, scientists have been surprised to discover dinosaur fossils in Alaska. Dinosaurs need warm temperatures—warmer than Alaska was thought to have been. In 1986, on Canada's Axel Heiberg Island, only 600 miles from the North Pole, scientists found a huge, dense forest. Some of the trees were well over 125 feet tall. They have been snapped off at the stump and are neither fossilized nor petrified, but remain as fresh wood. Even some of the leaves are still intact. The area was filled with spruce, hickory, dawn redwood and cypress trees and was evidently a swamp, since tropical turtle and alligator remains have also been found. Tree rings show that the area was warmer in the recent past, while the size of the rings tell us that there were short, hot summers during which rapid growth occurred, followed by long, dark winters as there are today.

The most interesting conclusion reached by the scientific team is that the entire region must have been subjected to a huge flood! This discovery tells us that all of geology must be restructured to present a history that agrees with the Bible's story of the earth's past.

Prayer: Dear heavenly Father, I thank You that Your Word is true and trustworthy. Let my faith be more than a faith of words or simply comfort in bad times. Instruct me through the Bible so that my faith is a way of life. In Jesus' Name. Amen.

REF: Search for a Tropical Arctic (Television documentary). 1990. Federal Department of Energy, Mines and Resources (Canada).

The Never-Fail Bat Alarm

Psalm 119:154

"Plead my cause, and deliver me: quicken me according to thy word."

The swiftly moving bat cuts almost invisibly through the darkness, silent to all but those who can hear its ultrasonic radar. What chance does a poor night-flying cricket—prized as a tasty bat snack—have?

Let's treat this problem as if *you* were the Creator, faced with the problem of giving your night-flying cricket a chance against its predator. Of course, you could design the cricket so that it would stay in its hiding place at night, but then you wouldn't have a night-flying cricket. You could give the cricket the ability to hear the bat's radar, but that alone won't help the cricket if he can't get to safety.

The solution our true Creator has invented to help the cricket preserve its life is another example of His engineering excellence. He has built the cricket with a single-celled bat detector that is wired into the cricket's nervous system. The detector is triggered by the frequency that bats use for navigation. When triggered, the cell fires off up to 500 impulses per second. These impulses automatically cause the cricket to steer away from the source of the ultrasound. The most impressive engineering fact about this bat detector is that it only works if the cricket is in flight and therefore vulnerable to bats. When crickets are safe from bats—resting, hiding, eating or preening—the bat detector cell doesn't fire any warning at all!

The precise elegance that the Creator has designed into the cricket for its protection is a witness to His wisdom in caring for all that He has created. Learn more about that wisdom in the Bible.

Prayer: Dear Father, Your love is so great that it extends to all that You have made. Your love for me is deeper than I can ever know, but I thank You that I can know that You so loved me that You sent Your only Son for my salvation. Thank You for Your love. Help me to reflect that love to others. Amen.

REF. Shreeve, James. 1985. The great cricket escape. *Science 85*, May. p. 83.

An Internal Clock You Can Feel

Acts 17:25

"Neither is worshiped with men's hands, as though he needed any thing, seeing he giveth to all life, and breath, and all things;"

We all have a number of clocks within us that help keep life constant in changing surroundings. The pineal gland is one clock that helps us adjust to our surroundings like the changing lengths of daylight through the year.

Earlier in this century, many scientists, having no understanding of the pineal, decided that it had no purpose and was left over as a vestige from our evolution. Creationists knew it must have a purpose because they knew we have not evolved.

If you are one of those people who typically feel a bit more lethargic or even depressed during the darkest months of winter, you might be able to blame your pineal, which is located near the center of your brain. How can a gland deep in the center of the brain know how long the days are? Light causes our bodies to make certain chemicals. When days are short, those chemicals are scarce, so the pineal cannot make from these chemicals the hormones that encourage the alertness we feel in the spring and summer. In addition, your pineal has a private nerve hookup to your eyes, so it can literally see for itself how much light is available! Scientists have found that adding artificial light during winter can sometimes help the winter blues.

Our bodies are a network of complex control systems and chemical factories—a complex union of systems that go beyond the abilities of the best human designers, but which testify to the wisdom of the Creator!

Prayer: Dear Father, if I had to design and build my own body, I would never have walked this earth. Help me to remember that just as You have given me this material life, so I must also look to You alone for eternal life, which You give me through the forgiveness of sins, earned on the cross for me by Your Son, Jesus Christ. Amen.

REF. Fellman, Bruce. 1985. A clockwork gland. *Science 85*, May. p. 77.

Creation Scientist Cleans Up the World!

John 3:12

"If I have told you earthly things, and ye believe not, how shall ye believe, if I tell you of heavenly things?"

Much of the world's water pollution is the result of biological aging and overloading. If, for example, too many nutrients encourage too much growth in the upper layers of the lake, eventually the systems at the lake bottom that dispose of the dead creatures from the upper layers will overload. If that happens, lake water will have an accumulation of toxic gasses and less oxygen—a combination that can destroy the familiar and more popular game fish in the middle and upper levels of the lake.

When Bob Laing, a creationist, looked for a solution to this problem, the first question he asked was, "Did God design a solution to this problem?" The answer is found in the way lakes turn themselves upside down twice a year. In the spring and fall, as water temperatures change through 39° (F), the bottom waters are pushed to the surface by heavier, 39° water. At the surface these "bottom waters" give up their poisonous gasses and absorb oxygen. Laing went on to prove, using huge air bubbles and other equipment, that lakes could be saved by stopping the original source of the problem and then artificially duplicating the Creator's built-in system.

Environmentalists at first rejected his solution. However, today, Bob Laing and his Clean-Flo Laboratories, has restored over 1,000 lakes and ponds worldwide, including the Tokyo Garden Ponds, where each fish is worth up to $100,000!

Bob Laing arrived at his revolutionary approach to solving lake pollution *because* he is a creationist. He knows that that creation is intelligently designed by our loving Creator.

Prayer: Dear Lord, because all things have been created through You, Your solutions are always the best. Forgive me for thinking that You are interested only in the spiritual and seeking my own wisdom and solutions to worldly problems. Amen.

REF: Runholt, Susan. 1990. Clean-Flo Laboratories . . . cleaning up the world. *Corporate Ink West* (West Suburban Business Communities, Minneapolis, MN). Sept. p. 1.

Fields and Forests of Medicine

1 Peter 1:20-21

"Who verily was foreordained before the foundation of the world, but was manifest in these last times for you, Who by him do believe in God, that raised him up from the dead, and gave him glory; that your faith and hope might be in God."

Blessings can become curses. For example, a painkiller may be used to cover up an injury that needs immediate treatment. Pain is unpleasant. However, when pain tells us to move our hand from the hot stovetop, it helps us prevent worse injury.

When God created the universe and placed humans on this earth, there was no disease, no pain, and no death. The Bible tells us that God also knew that our freedom to withdraw our love from Him would also result in the consequences of withdrawing that love—that is, disease, pain and death. In His wisdom, God filled the creation with chemical factories that could provide temporary help for pain and disease.

Foxglove, or digitalis, offers a treatment for heart failure. Quinine, from a South American tree, is used to treat malaria. Steroids come from a Mexican yam. A treatment for gout is made from the autumn crocus. A fungus that grows on the rye plant and caffeine from coffee beans are used to treat migraine headaches. White willow bark, which contains a form of aspirin, has been used for thousands of years as a natural treatment for pain and fever.

Pain, disease and death call our attention to our much more serious spiritual condition. If we did not have to face death in this life, we would be condemned to spend eternity in an imperfect world. We can thank God that He has provided the remedy for our sin in the forgiveness of sins through His Son, Jesus Christ. Is your trust for salvation in Jesus Christ and what He has done for you?

Prayer: Dear heavenly Father, I thank You that You have provided us with relief from some of the earthly results of sin. Do not allow earthly remedies to hide my need for the only true remedy for sin, which is found in Jesus Christ, Your Son. Amen.

REF.: Many medicines have plants as their source. Minneapolis *Star-Tribune*, Nov. 1, 1990. p. 11E.

Ancient Well Surprises Scientists

Romans 1:22-23

"Professing themselves to be wise, they became fools, and changed the glory of the uncorruptible God into an image made like to corruptible man, and to birds, and fourfooted beasts and creeping things."

A few years ago, archaeologists discovered a water well that is the oldest they've ever seen. Because of the well's sophisticated design, they immediately concluded that ancient humans were much smarter than they ever expected.

The well was found beneath the Mediterranean Sea bed off the coast of Israel. The location in which the well was discovered is part of a 10-acre site of an ancient city that was found beneath 10 feet of sand and 39 feet of seawater. Other more shallow wells have also been found at the site as well as stone houses, fireplaces, 23 skeletons, seeds, tools—and the village garbage pit.

The deepest well is a round shaft, extending 16 feet into the ground and paved with stone. Evolutionary archaeologists said that the discovery raises some interesting questions. For example, how did ancient humans know that if they dug deep enough they would find the water table? Archaeologists also wondered where ancient humans learned the skill of sinking a round shaft and paving it with stone.

Finds like these, which extend so-called modern humanity with all their skills and intelligence further into the past, are no surprise to those who accept the Bible's account of human history. We would guess that this city was built within several generations after the great Flood. That date is not ancient for human beings, since we have been fully human and skilled since the sixth day of the universe's existence. In fact, based on the Bible's description of Adam and the first generations of humans, we are probably less skilled today than most ancient people.

Prayer: Dear Father in heaven, I thank You that You have created humankind with which to share Your love. Help me to put away those values that the world holds because it thinks people came from animals and adopt those values that come from You because You made us. In Jesus' Name. Amen.

REF.: Ancient well offers clues about early people. *Wichita Eagle*, Sept. 28, 1990.

The Underrated Sense

Romans 8:29

"For whom he did foreknow, he also did predestinate to be conformed to the image of his Son, that he might be the firstborn among many brethren."

How does your nose work? Medical researchers admit that they don't know. Your nose performs an extremely unlikely job. It senses invisible molecules in the air and turns each molecule into a unique sensation that is sent to the brain. If the fact that molecules floating in the air could be detected was a modern discovery, it would be hailed as the scientific discovery of the century!

Medical researchers do know that the roof of your nose cavity contains about 10 million cells that are wired directly into the brain. The nervous system that connects your olfactory cells to your brain is so large and complex that its surface area is several times greater than the surface area of your body. Researchers also know that your sense of smell is important to your sense of taste. For example, that's why, when you have a head cold, chocolate tastes like a flavorless waxy paste. In this case what you taste in your mouth has more to do with what is happening in your nose than on your tongue.

Many creatures, including birds, can sense odors. Farmers who had difficulty with birds eating as much as 20% of the food set out in feed lots discovered that birds hate the smell of grapes. When they laced the feed with grape scent, the birds left it alone, but the cattle didn't seem to mind.

Researchers have also learned that each of us has a unique "odor print." This reminds us that each of us is a unique being, made by our Creator because He *wanted* to have a relationship with us.

Prayer: Dear Lord, I thank You for the richness that the gift of smell adds to my life. Your provision of such additional rich experiences reminds me that You are concerned about my life both here and in eternity. In Jesus' Name. Amen.

REF.: Dawson, Jim. 1990. Nose is now. Minneapolis *Star-Tribune*, Nov. 8. p. 20A.

The Orchid that Captures Workers

Job 9:10

"Which doeth great things past finding out; yea, and, wonders without number."

The relatively rare bucket orchid, *Coryanthes*, behaves almost like an animal at pollination time. This remarkable ability is essential to its survival.

Coryanthes has a steep-sided flower. Two glands extend over the center of the "bucket" and secrete a clear fluid into the flower after it opens. Just above the pool of fluid inside the bucket, a tunnel opens to the outside of the bucket. At the end of the tunnel are the flower's pollen and stigma.

When it opens, the flower sends out a strong sweet odor that can attract male bees from over five miles away. The bees collect a waxy material on the flower's surface that they later use in mating rituals. As the number of bees collecting this substance off the flower's surface increases, so does the likelihood that in the excitement one of them will fall into the pool below. When this happens, the sticky fluid makes it impossible for the bee to make it out of the "top" of the bucket-shaped flower. However, the tunnel provides an easy exit. But as the bee nears the tunnel's end, the flower drops down a projection from the tunnel's ceiling, holding the bee for about ten minutes before freeing him. While the bee is held, the flower glues two packets of pollen to the bee. If it should happen that the bee already has pollen packets, this activity delivers the pollen to the stigma, and pollination is complete.

The *Coryanthes* orchid not only shows us the Creator, but it shows us how unlimited His creativity and imagination are.

Prayer: Dear Father, no human being could plan and design the wonders we see all around us. I ask that more of us who know You and Your love for us in Jesus Christ would be bold to add our words to the witness of Your truth, so that others may not only know that You are, but that You offer us forgiveness in Your Son, Jesus Christ. Amen.

You *Are* Left-Handed

1 John 4:19

"We love him, because he first loved us."

If I were to ask you whether you are left-handed or right-handed, you would probably answer with most people that you are right-handed. While that may be true, you are also left-handed, and your life depends on it.

Biologists have absolutely no idea why molecules of living things are left-handed. You see, when biological molecules are made in "origin of life" experiments, they are always a mixture of both left-handed and right-handed molecules—mirror images of each other that won't work together in living things. A right-handed amino acid can be chemically identical to a left-handed amino acid. But for some mysterious reason, a living organism will treat the right-handed amino acid as if it were useless, even though normal chemical reactions ignore the difference. This tells us that living things are much more that just the result of chemical reactions. As the Bible suggests when it describes the creation of life, living material is different in some very important ways from nonliving material.

The fact that life uses only left-handed versions of biological molecules poses another problem for those who think that life formed spontaneously, without a Creator. While experiments show that some simple biological molecules can be formed in the laboratory, experiments always produce a mixture of left-and right-handed molecules.

So life's origin, with its left-handed-only molecules, remains a mystery to those who reject the Creator. Could this be why God designed life with left-handed only molecules in the first place?

Prayer: Lord, not only excellence of design, but also wisdom crowns Your work of creation as You seek to gain the attention of those who are running from You. I thank You that Your love did not allow You to give up on me. Amen.

REF: Olson, Steve. 1984. Why is life handed? *Science 84*. p. 22.

Changing Poison into Medicine

Genesis 3:18

"Thorns also and thistle shall it bring forth to thee, and thou shall eat the herb of the field."

If God created the world and everything in it in perfection, where did disease, sickness and death come from? It's clear from the Bible that God did not intend for humans to suffer disease and death. However, when humans withdrew their love from God by disobeying Him, they caused the nature of creation to be changed.

It appears that it would take only small, subtle changes in chemistry and biology to change helpful and good organisms into deadly organisms. For example, botulism is a fatal food poisoning caused by the botulinum toxin produced by a microbe growing in food. Botulinum is the most powerful toxin known to modern science.

Medical researchers have discovered that very tiny amounts of botulinum toxin may be safely injected into spasming muscles to stop involuntary spasms. These small doses stop involuntary nerve impulses without interfering with voluntary use of the muscles. This therapy promises to help people who suffer from involuntary clinching of eyelids, throat, misaligned eyes, stuttering and the cramps often suffered by musicians, writers and typists. Therapy appears to work for several months before it must be repeated, and there appear to be no side effects. This treatment is not theoretical; doctors are being trained in using it right now.

So a dangerous poison *can* be medicine. Perhaps this hints at a purpose God may have originally intended for the botulism microbe.

Prayer: Dear Father, I know that human sin has ruined the creation. I confess that I, too, have sinned and have not helped Your creation. I trust Your promise of the forgiveness of sins through Jesus Christ for my salvation and a restored relationship with You. In Jesus' Name. Amen.

REF.: Toxin treatment helps variety of muscle spasms, panel says. Minneapolis *Star Tribune*, Nov. 15, 1990. p. 10A.

Learning in the Womb; Learning from the Womb

Job 10:10-11

"Hast thou not poured me out as milk, and curdled me like cheese? Thou hast clothed me with skin and flesh, and hast fenced me with bones and sinews."

January 22, 1973 marks the disastrous turn in American law that granted women the "right" to kill their unborn babies. The Bible makes it abundantly clear that the unborn child is a human being who is known and loved by God as He knits it together in the womb. Science knows that even when it is a single fertilized cell, the unborn child is uniquely human. Now science is learning the truth of what the Bible has told us all along about the individual human nature of the unborn child.

Research is confirming that even before birth children in the womb are learning about the world. The child in the womb does much more than simply hear sounds, including the sound of its mother's voice. The child also becomes familiar with odors in the outside environment.

To test what newborns were thinking, scientists cleverly hooked up a pacifier to a tape recorder. They then rewarded a particular sucking pattern with various sounds such as mother's voice. What they found was that the infants would concentrate on a pattern that would produce their mother's voice. In another test, women's voices, and even father's voices, were not preferred over mother's heartbeat—a familiar sound in the womb.

Even science is beginning to agree that life in the womb is nothing less than a normal stage of human existence. Like every other stage of life, life in the womb includes learning new things and developing relationships. If only this knowledge was being applied to the abortion debate.

Prayer: Dear Father, I ask that You would not turn Your face from us because of the millions of infants killed by abortion. Millions of Your people reject abortion. Grant us strength to uphold Your truth wisely and end the deaths. In Jesus' Name. Amen.

REF.: Recollections from the womb. *Science 84*, Dec. 1984. p. 84.

The Mystery of Male and Female

Genesis 5:2

"Male and female created he them; and blessed them, and called their name Adam, in the day when they were created."

"God created them male and female." That simple phrase heaps a world of problems on those who believe in evolution.

Evolutionists tell us that in the early stages, evolution would have produced creatures that reproduce without the need for male and female. They then suppose that some mutation would have produced the gender differences, but what advantage would these differences offer? Quite the contrary, these differences would be a disadvantage and become quickly selected out. An even more serious problem is that, according to evolution's history of life, the male and female sexes must have selected in different kinds of creatures *dozens* of times.

No one has the slightest idea how one or a dozen mutations could have produced male and female. To change a creature from having no gender into a male or a female requires developing several new organs, changing some others, creating entire new hormone systems, and changing others. New organs and systems would have to be immediately integrated into the workings of the rest of the body. In addition, the independently developed male changes would have to work with the independently developed female changes. Changes would also have to be made in the brains of the new creatures so that they would want to mate with each other. And evolutionists tell us this must have happened dozens of times!

Yes, that simple phrase "God created them male and female" leaves evolution with an imponderable mystery. This mystery is solved easily by God's Word. What a mistake it is to say that the Bible is obsolete in our modern scientific era!

Prayer: I thank You, Lord, for the gift of sex. Renew my heart and mind so that I desire to use this wonderful gift in the way You intended. Amen.

REF.: Ocvirek, Steven. 1990. The gender gap. *Bible-Science Newsletter*, Dec. p. 5.

Simple Math, Hard Questions

Isaiah 40:21

"Have ye not known? have ye not heard? hath it not been told you from the beginning? have ye not understood from the foundations of the earth?"

Today the Bible is often accused of giving answers that don't add up to reality as we know it. Many Christian students have been told that evolution gives answers that do make sense. Is this really true? These are questions that we Christians can and should ask when we are told that evolution fits the world as we know it.

Evolutionary geology says that, on the average, geological processes such as erosion and mountain building have always taken place at the same rate we see them happening today. However, if we take the volume of the world's continents and divide it by the known rate of continental erosion, we find that it should take only 14 million years to erode the continents down to sea level. We are also told that the last mountain building activity took place about 65 million years ago. So if evolutionary geology is correct, the continents should have long ago been worn down to sea level.

Evolutionary biologists tell us that the first modern women lived about 200,000 years ago. At an average of three children per family (the current world average is four), the world's population should have reached 3.5 billion in 4,000 years. In fact, we reached that figure only within recent centuries. Yes, wars and disease accounted for more losses in the past, but this fact does not explain this huge error.

More examples could be mentioned. The truth is, the facts don't add up to evolution. They *do* support the Bible's view of history. Explanations that deny the Creator just don't add up.

Prayer: Lord, I thank You that You have filled the creation with evidences that make it difficult to deny You. Help me to use these as a witness to bring people to Your Word where they learn of the forgiveness of sins. Amen.

REF: Dye, Brad. Evolution vs. the big picture. *Dialogue*.

Did a Woman Write Genesis?

1 Corinthians 8:4

"As concerning therefore the eating of those things that are offered in sacrifice unto idols, we know that an idol is nothing in the world, and that there is none other God but one."

For thousands of years, theories that reject the Word of God have followed cultural fashion. It was inevitable, then, that some "biblical scholar" would invent a theory that a woman wrote some important part of Scripture that has been traditionally assigned to a man.

Historical critic Harold Bloom, who teaches at Yale, has proposed that the mythological "J" source of Genesis was a woman. He says she made up the stories of the creation, Noah and Abraham, and invented an impulsive, violent and "impish" divinity. Bloom's evidence? The account of Eve's creation is six times longer than the account of Adam's creation.

Some Christians have been shaken up needlessly by this so-called "scholarship." Historical criticism rejects the Bible as the Word of God. It says the Bible is man-made and a result of the natural evolution of religion. There is no room in historical criticism for divine revelation. As a result, historical critics cannot arrive at theories that support the Bible as God's revelation except by accident. Second, the claimed sources of the Old Testament, such as the "J" source, have never been found and are purely imaginary.

As Christians we should not be intimidated by such claims of "scholarship." All this scholarship does is discard the God of Scripture and His Word, replacing the true God with imaginary gods and authors. And as we read in 1 Corinthians chapter 8, "we know that an idol is nothing in the world, and that there is no other God but one."

Prayer: Dear Lord, You are the Word made flesh for our salvation. You are the message of all of Scripture. Grant us teachers who are faithful to Your Word and love to teach the truth. Amen.

REF.: Waddle, Ray. 1990. Is Genesis fiction by female writer? *The Tennessean*, Oct. 7. p. 20.

Wild Medicine

John 14:6

"Jesus saith unto him, I am the way, the truth, and the life; no man cometh unto the Father, but by me."

Is it possible that your pharmacist could be a chimpanzee? That's not such a silly question. Naturalists are discovering that many animals use a variety of plants and other items to treat their illnesses. Animals often know about the medicinal properties of plants when modern medicine has yet to learn about them.

Recently naturalists noticed that one of the monkeys in the African preserve under their care had become lethargic, lost her appetite, and obviously had a severe gastric upset. A few hours after she became sick, she started munching on a native shrub that is not a normal part of the monkey's diet. She chewed the plant, swallowed the bitter juice and spit out the pulp. By the next afternoon her symptoms were gone and she was feeling better. After a little research, scientists discovered that local tribes use juices from the same plant to treat intestinal upset and loss of appetite.

Studies of the juices of a root that bears rub into their coats show that the bears are using an effective insecticide. Scientists have observed elephants, monkeys, birds and bears eating a variety of things that were not a normal part of their diet only to discover that the things being eaten had medicinal properties. Most amazing is that the animals know how best to take their medicine. If a leaf is best swallowed whole to kill intestinal parasites, it will be swallowed unchewed!

Who other than the Great Physician Himself could have taught the animals this medicine?

Prayer: Lord, You are the Great Physician of body and soul. I thank You that You have enabled the animals to find relief from disease. Teach me to consider my soul as well as my body when I think of good health for myself so that I seek that healing and health that only You can provide. Amen.

REF.: Cowen, Ron. 1990. Medicine on the wild side. *Science News*, v. 138. p. 280.

Spiders Help Make the Sun

Proverbs 30:28

"The spider taketh hold with her hands, and is in kings' palaces.

The instruments that scientists are using that develop fusion energy continue to become more complex and powerful. They are trying to generate the energies needed to cause hydrogen to fuse into helium. This is the same process thought to be going on in the sun. Once the fusion process is started, it will continue on its own as long as hydrogen is available, quickly generating much more energy than was used to start the process. Hydrogen fuel is easily and cheaply generated from seawater.

Scientists have been using lasers to generate the power needed to start the fusion process. In current experimental devices, the hydrogen fuel is held in a hollow glass sphere that is about as big around as a human hair. The sphere must be suspended in mid-air. To reach the 90 million degrees needed to start fusion, 24 lasers with 12-trillion-watt beams are aimed at the sphere.

What do you suppose scientists used to suspend the sphere? They thought of using strands of glass, but then they remembered something even stronger, yet thinner—spider silk. Spider silk has the added benefit of being sticky, so it easily holds the sphere in place.

Despite the impressive technology of fusion experiments, humans often find that they cannot create better materials than those found in nature. God freely gave His creatures a sophisticated array of materials for life. He has just as freely given His Son, Jesus Christ, into death for the forgiveness of your sins.

Prayer: Dear Father, I thank You for Your generosity in all things. I especially thank You that You did not even withhold the life of Your only Son for my salvation. Teach me such generous love. In Jesus' Name. Amen.

REF: Arachnid fusion: hanging by a thread. *Science 84*, Nov. p. 15.

Plastic Stone and Other Ancient Wonders

Ecclesiastes 1:9

"The thing that hath been, it is that which shall be; and that which is done is that which shall be done: and there is no new thing under the sun."

It is unfortunate that we so often have the sense that nothing technologically important has happened in human history until modern times. This helps maintain the illusion that humans are indeed progressing and evolving upward.

Differential gears, allowing a machine to perform two functions simultaneously, were not used in modern times until 1575. However, a model planetarium using differential gears was found in a ship that sank off Greece in 78 BC. While Ben Franklin is given credit for discovering electricity, electric batteries dating from 100 BC have been found in Iraq.

The Gate of the Sun is a ten-ton intricately carved stone gate in Bolivia, 13,000 feet above sea level. In Mexico there are boulders with faces carved on them. They were erected between 1250 and 400 BC, carved out of boulders weighing up to 24 tons that were quarried 45 miles away from their present location. How were these giants moved and erected? Joseph Davidovits, a scientist at the Geopolymer Institute in France, believes he may know. He suggests that the builders chemically broke down the stone with acid. Then they carried the liquid stone to the present sites and poured the liquid into molds. In support of his theory, Davidovits points out that some Indians still use this method now, while traces of the acid can still be found in some ancient stone structures.

Despite all of our modern gadgets, there is no evidence that humans are any more clever today than they were on the first day they walked the earth—just as the Bible says.

Prayer: Dear heavenly Father, humans easily fill themselves with pride over their accomplishments. I confess that my pride, too, gets out of control. Forgive me for the sake of the innocent suffering and death of Your Son for me. Teach me how to better glorify You rather than myself. In Jesus' Name. Amen.

REF.: Patton, Robert. 1982. Ooparts. *Omni*, v. 4. p. 53.

Is There a Gap in Genesis One?

1 Corinthians 15:22

"For as in Adam all die, even so in Christ all shall be made alive."

One of the most frequently asked questions when I appear on a radio phone-in program is whether there is a gap between the second and third verses of Genesis chapter 1. Some people have been taught that a gap of millions of years between the original creation of the earth and the earth we know today can explain the fossil record and the millions of years proposed by evolution.

Also called the ruin-reconstruction theory, this idea says that an earlier creation existed that was judged by God. Those who believe in this theory usually place Satan's fall, dinosaurs, and so-called "cave men" into this earlier creation.

The gap theory was first proposed in the last century by Rev. Thomas Chalmers as a response to the growing popularity of long evolutionary ages. It was widely spread among Christians in the notes that first appeared in the *Scofield Reference Bible* of 1917.

The gap theory, however, does not satisfy evolutionists, nor does the Hebrew of Genesis support it. Most importantly, the Bible teaches us that death first came into the world when humans began to sin. The Bible repeats many times that sin and death began with Adam and not before. This is such an important point that the Bible links the beginning of sin and death with the first Adam and the victory over sin and death with Jesus Christ, spoken of as the Second Adam.

Of course, there is much more to the issue. For further information, why not call Creation Moments and ask for literature on the gap theory.

Prayer: Dear Lord Jesus Christ, I offer You grateful thanksgiving because You became the Second Adam in order to save us from sin, death and the devil, our lot because of our sin. Amen.

REF.: Niessen, R., B. Northrup, and D. Watson. 1988. *Genesis Stands*. Minneapolis, MN: Bible-Science Association, Inc.

The Well-Designed Living Fossil

1 John 4:6

"We are of God: he that knoweth God heareth us; he that is not of God heareth not us. Hereby know we the spirit of truth, and the spirit of error."

The *coelacanth* is a rather strange looking fish that was supposed to have become extinct 70 million years ago with the dinosaurs. That's what scientists thought anyway. After all, they argued, the fish looks so primitive. Its fleshy fins are obviously the first step in the evolution of legs. Meanwhile, fishermen off the east coast of Africa were familiar with the creatures because the fish were turning up in their nets.

Science discovered in 1938 what these fishermen already knew. In the last half-century more than 80 of the creatures have been caught and studied. Deep-diving submarines have also allowed scientists to study the *coelacanth* in its natural habitat.

As science has learned more about the *coelacanth*, it has become clear that this fish is not primitive, nor does it provide aid and comfort to evolution. Studies on living creatures show that the fins are structured all wrong if they are evolving into legs. They're just fins. In addition, the *coelacanth* is a remarkably sophisticated creature. It reproduces more like a mammal than a fish, giving birth to live young after about a year's gestation. The *coelacanth* has a small second tail that improves its swimming abilities. It also appears to have a gland that gives it the added ability to find its prey by detecting the prey's electrical signals.

Science had declared the *coelacanth* extinct, primitive, and an evidence for evolution. Now having studied them, we learn that each scientific conclusion was wrong and that the *coelacanth* glorified its creator by its sophisticated design.

Prayer: Dear Father, grant that we would seek the wisdom of Your Word and receive a strong faith through the working of Your Spirit so that we are not easily led into error by the unbelieving world. In Jesus' Name. Amen.

REF.: No stinking fish in my taxi! *Discover*, Mar. 1985. p. 40.

A Truly Strange Bird

Psalm 40:5

"Many, O LORD my God, are thy wonderful works which thou hast done, and thy thoughts which are to us-ward: they cannot be reckoned up in order unto thee: if I would declare and speak of them, they are more than can be numbered."

Though God created the entire living kingdom in only a few days, the variety and creativity of what He made seems nearly unlimited by our standards. One of the more unusual creatures He made was thought extinct until it was rediscovered in 1958.

The kakapo parrot lives in New Zealand. The most unusual parrot on earth, it is one of only a few known parrots that prefers to sleep during the day and becomes active at night. Weighing in at five pounds, it is also the world's heaviest parrot. It is, perhaps not surprisingly, the world's only non-flying parrot.

The Creator's unusual expression of inventive creativity in designing the kakapo did not end here. The mating habits of the kakapo are especially peculiar for birds. In mating season the males gather in locations that are used year after year for mate selection. Female parrots come to these places to inspect the males to select a mate. However, in most un-bird-like fashion, the males provide absolutely no help building the nest or rearing the young.

The kakapo is remarkable because of its many strange traits, most of which would make it least fit for survival. In other words, not only is it an unusual creature, but its more unusual characteristics seem to put it at a disadvantage as far as evolution is concerned. So while evolution would not have made the kakapo, our inventive Creator did, perhaps as a witness against evolution.

Prayer: Dear Father in heaven, I thank You for the beauty and creativity You have placed into Your work of creation, and for the blessings these gifts add to life. In Jesus' Name. Amen.

REF.: *Discover*, Mar. 1985. p. 36.

A Modern Myth

2 Peter 1:16

"For we have not followed cunningly devised fables, when we made known unto you the power and coming of our Lord Jesus Christ, but were eyewitnesses of his majesty."

A growing number of scientists, some of them with worldwide reputations, are changing their minds about evolution. While many of these scientists are recognizing that evolution is in trouble, they are not necessarily becoming Christians or creationists.

The world famous astronomer Sir Fred Hoyle announced that there is so much information within higher forms of life that they could not have evolved. Dr. Hoyle said that he could not understand why so many biologists deny this obvious fact and accept evolution.

The scientific world was solidly shaken when Dr. Colin Patterson, head of paleontology at the British Museum, admitted at the American Museum of Natural History in Chicago that he knew *nothing* about evolution that was true. He said that he had asked other scientists if they could name one thing about evolution that was true. Silence was the usual answer. Dr. Patterson called the evolutionary theory "anti-knowledge." He especially faulted those who say evolution is a fact, yet admit they don't know how it works.

Evolutionary theory is the great modern creation myth. As a myth, it is no challenge to the Bible. It's even losing support in the scientific community. If you would like more information, call Creation Moments and ask for information about scientists turning from evolution.

Prayer: Dear Father, the glory of Your handiwork in the creation is so great that people have difficulty inventing explanations that deny Your work. Help me use this to witness the salvation that is found in Christ. Amen.

REF.: Sunderland, Luther. 1988. *Darwin's Enigma*. El Cajon, CA: Master Books.

Designed to Renew

2 Peter 3:10

"But the day of the Lord will come as a thief in the night; in the which the heavens shall pass away with a great noise, and the elements shall melt with fervent heat, the earth also and the works that are therein shall be burned up."

We live in a world—indeed a universe—that is in continual change. Things have been designed that way for a very important reason. Unfortunately, some people choose to define evolution as simply *change*. However, that's not the kind of continuous change that surrounds us.

The change we see around us serves to keep things the same. Biologists call this stasis. That might sound contradictory, but think about it for a moment. A baby is born—that's change. The baby grows into a child and then a young adult. That's more change. In the end, though, a new generation is born to replace the last generation, and things are as they were. It's the cycle of life, common to all living things.

What's more, the interrelated life cycles of all living things complement each other. Plants use sunlight and carbon dioxide to make oxygen and food, which are needed by the animal kingdom and human beings. Each generation of living things reproduces after its own kind, helping to keep this balance in tune.

The change we see all around us was designed by the Creator to renew the creation so that it might remain stable—the same. We can truly say that the more things change, the more they stay the same. However, this state of affairs will not go on forever. One day—and perhaps soon—the world will end. The Son of God who was the instrument of our creation and the instrument of our salvation will return. Are you ready to meet Him?

Prayer: Dear Lord, help me to be prepared to meet You without fear when You return. I trust in Your promise of salvation through the forgiveness of sin that You earned for me on the cross. Amen.

Not Always What They Seem

Matthew 10:29

"Are not two sparrows sold for a farthing? and one of them shall not fall on the ground without your Father."

The creation is filled with a remarkable number of creatures that are able to fool their enemies into thinking they are something else. They provide us with examples of God's attention to details in the creation.

The toad grasshopper of South Africa may be found among different kinds of rocks. Grasshoppers that live among quartzite pebbles have the same lighter colors and markings as the pebbles, making them nearly invisible. Across the valley, other toad grasshoppers living among the volcanic rocks are darkly colored like the volcanic rocks.

The Kenyan bush cricket lives almost invisibly among the moss that commonly covers trees, because it looks just like the moss. The Trinidadian katydid looks as if it might be a patch of blotchy fungus that grows on the same leaves it lives on. The Malaysian katydid takes a different approach—it looks just like one of the veined leaves upon which it lives. The Malaysian tussock moth has transparent patches on its wings so that with its coloration it looks like a decayed section of the leaf upon which it lives. The larva of the silver king shoemaker butterfly eats leaves in the protection of nights' darkness and by day hangs from the leaf it has been eating, looking like part of the tattered leaf.

No creature was too small or unimportant when our Creator planned the details of His work. We should see from these examples that He cares for each of us. His message of how He cares for *you* is in the Bible.

Prayer: Dear Father in heaven, I confess that I have often placed value on things in the same way that the unbelieving world does. Teach me in Your Word how and what You value. In Jesus' Name. Amen.

REF.: The masters of mimicry. *Discover*, June 1985. p. 56.

The Father Who's a Good Mother

Job 11:7

"Canst thou by searching find out God? canst thou find out the Almighty unto perfection?"

When it comes to designing the seahorse, God showed that there are no limits to His creativity.

There are about 35 known species of seahorse, the largest of which grow to over a foot in length. Seahorses are masters of disguise. They can change color in seconds to match new surroundings. While seahorses are fish, they have skin instead of scales. Their skin, which is tightly stretched over bony plates, produces growths to match surrounding vegetation. These growths will change to match new vegetation or disappear if there is no vegetation. Although seahorses have neither teeth nor stomach, they eat small, tough crustaceans. They are able to grasp so tightly with their prehensile tails that they cannot be removed unless they *want* to move. Seahorses are also very social creatures, greeting their neighbors each morning by changing colors.

The seahorse is the only animal known where the male actually becomes pregnant and gives birth. Normally the mating ritual lasts for three days. It includes color changes, dancing and grasping. At the end of the ritual the female deposits her eggs in the male's pouch. For the next two weeks the male protects, nourishes and aerates the eggs. When they are ready, the young are born alive and ready to swim. One male, whose pouch was a little bigger than half a teaspoon, gave birth to 1,572 young!

By its very existence the seahorse offers elegant witness to a powerful, wise and personal Creator.

Prayer: Dear Father, I shall never understand all that You have made, it is so great and wonderful! Let my sense of appreciation grow and move me to tell others of Your greatness and Your love for us in Jesus Christ. Amen.

REF.: Vincent, Amanda. 1990. A seahorse father makes a good mother. *Natural History*, Dec. p. 34.

Heating for Health

Proverbs 3:7-8

"Be not wise in thine own eyes; fear the LORD, and depart from evil. It shall be health to thy navel, and marrow to thy bones."

You're not feeling well and you have a fever. Is your increased body temperature good for you or should you try to lower it?

Medical researchers are learning what happens in the body when we develop a fever. They say that the body's fever response is usually a very important part of our body's defense against disease. Normally the body's temperature varies through the day. If the body's temperature drops below 80° or climbs higher than 108° for a prolonged period, death usually results. Your body has a thermostat, called the hypothalamus, to keep its temperature within that range.

When disease germs are detected in your bloodstream, your white blood cells release a chemical called EP. EP quickly makes its way to the hypothalamus, where it raises the setting of your thermostat. This increase in temperature can make your body increase its production of T cells by 20 times or more to fight bacterial infection. It also increases the production of blood chemicals that fight viruses. EP also blocks bacteria's ability to make use of the free iron in your blood at just the time bacteria are most in need of that iron. Even iguanas and fish are known to raise their body temperature when ill by moving into warmer environments.

Does the fever response just *accidentally* happen to help fight infection, or was it *designed* to work the way it does? Clearly we must thank God for this cleverly designed defense against disease.

Prayer: Dear heavenly Father, I thank You for the good health that I enjoy as well as the wonderful ways You have designed the body to protect itself from disease. In Jesus' Name. Amen.

REF.: Kiester, Edwin Jr. 1984. A little fever is good for you. *Science 84*, Nov. p. 168.

Genesis, Naturally

Psalm 119:140

"Thy word is very pure; therefore thy servant loveth it."

Many people today think it's unusual for someone to interpret the Genesis account of creation literally. For example, many in the church seem unsure of whether Adam was a real person. How are we to understand Genesis?

Until the last century, virtually all Christians understood Genesis chapter 1 as a history of what actually happened during creation week. The few exceptions were influenced by Greek and Roman philosophy that was very much like modern evolution. For thousands of years believers understood the text of Genesis chapter 1 in its natural sense. In other words, unless the Bible itself gives us some reason to do otherwise, we understand the words in their literal, natural sense.

We could mention many leaders of the church who understood Genesis creation as a historical account of actual events. Most importantly, Jesus Christ treated the Genesis account of creation as actual history and Adam and Eve as real people. In defending the natural understanding of Genesis chapter 1, the great teachers of the church have most often cited Christ as the best example of how to understand Genesis.

The subject is, of course, much larger than can be covered in a few minutes. If you would like to know more about how to understand Genesis chapter 1, call Creation Moments to request information on understanding Genesis. We need to understand Genesis in its natural sense, of course, just as Jesus understood it.

Prayer: Dear Lord, I thank You that Your Word is so clear that even a child can understand what You are saying in Genesis, and yet a believing scientist can appreciate Your precision. I believe, Lord. Help my unbelief. Amen.

REF.: Niessen, R., B. Northrup, and D. Watson. 1988. *Genesis Stands*. Minneapolis, MN: Bible-Science Association, Inc.

Creative Strategies for Life

1 Corinthians 6:20

"For ye are bought with a price; therefore glorify God in your body, and in your spirit, which are God's."

Some creatures seem more wise and powerful in dealing with the problems of living than seems reasonable.

For example, when threatened with parasites, some creatures change their method of reproduction to put the parasites at a disadvantage. Several types of fungus infect some grasses, producing poisons. However, these poisons don't harm the infected grass. Instead, they protect the grass from insects and other enemies. This makes it less likely that the fungus will lose its happy home.

When fungal parasites take over certain wild grasses, the fungus destroys the flowers of the plant so that it can only reproduce by sending out shoots from its roots. Can you figure out the fungus's clever strategy? New shoots are genetically identical to the parent, and just as susceptible to the fungus. New plants from seeds, however, might express traits that were present, though unexpressed, in the parent, like resistance to the fungus. So it's in the fungus's best interest to restrict the grass to identical offspring.

A New Zealand freshwater snail can reproduce with or without fertilized eggs. However, in those lakes where the snail is threatened by a parasitic worm, the snail uses only fertilized eggs, increasing the likelihood that some of the offspring will be able to resist the parasite.

There is a lot more thought and power to be seen in these examples than worms, snails and grasses could generate. The Creator is even glorified in the relationships between parasite and host!

Prayer: Dear Father in heaven, there is nothing in the creation that does not glorify You! I thank You for that and pray that I might do a better job of glorifying You before others. In Jesus' Name. Amen.

REF.: Cowen, Ron. 1990. Parasite power. *Science News*, v. 138, Sept. 28. p. 200.

Tree Wars

1 Peter 5:6-7

"Humble yourselves therefore under the mighty hand of God, that he may exalt you in due time: Casting all your care upon him; for he careth for you."

A gypsy moth caterpillar infestation can strip the leaves from large numbers of trees, threatening the trees' lives. Since they don't have a means of escape, it doesn't seem as if the poor trees have much of a chance.

But botanists were tipped off to the possibility that trees may not be all that defenseless when they began to investigate why only a few trees were badly damaged by insects in an infected grove while most stood unharmed. They discovered that when facing a threat like gypsy moth caterpillars, trees begin to defend themselves by sounding the alarm. A large variety of trees —including beeches, poplars, sugar maples, and red oaks— communicate with each other. Scientists believe the trees communicate by releasing chemicals called pheromones into the air. What they now know is that before insects attacking one tree can get to the tree next door, the second tree has already begun defending itself.

How do they defend themselves? When under attack or notified by other trees of attack, most trees begin to manufacture an array of poisons. Some of the poisons make leaves impossible to digest while others kill the insects outright. Some make as many as eight poisons at once, and many can change the poisons that are made from year to year.

Our Creator's concern for living things extends even to plants. If He sees to this detailed level of care for trees, can there be *anything* in our lives that is too small for His loving concern?

Prayer: Lord, forgive me for those times when I have thought that a problem or concern of mine was too small to bring to You. Continue to urge me to become closer to You and more dependent on You. Amen.

REF.: Brownlee, Shannon. 1983. The silent battle. *Discover*, Sept. p. 28.

Musical Memory

2 Chronicles 7:6a

"And the priests waited on their offices: the Levites also with instruments of musick of the LORD, which David the king had made to praise the LORD, because his mercy endureth forever, when David praised by their ministry;"

Are you a musician? From the standpoint of evolution, human musical ability serves no purpose. Assuming that the laws of nature could even invent musical ability, they would have no reason to do so. However, from the standpoint of creation, music serves to praise our Creator.

Unfortunately, some of us who enjoy music cannot carry a tune, while other people can play or sing a piece without even looking at any music after hearing it once. The scientific study of musical ability has become an important area of study as a result of some recent discoveries. Researchers have discovered that it's true that some people have natural musical ability and some don't. They have found that some people have perfect pitch, even if they have no musical training, while those who are tone-deaf remain tone-deaf despite musical training. Studies show that only about one person in 2,000 has perfect pitch.

Researchers were shocked to find solid evidence that musical ability is genetically inherited. Studying musical ability in both musically trained and untrained families, they found that musical ability follows a dominant inheritance pattern. Sons or daughters have a 50% chance of inheriting the gene from a parent who carries it.

How could mindless evolution develop a gene that codes for musical ability? Even evolutionists appear to be puzzled by this. But those of us who believe in the Creator praise Him whether we have musical ability or not!

Prayer: Dear Father, I do indeed praise You for the wonder of Your creation. But most especially do I praise You for Your love toward me in sending Your Son, Jesus Christ, to save me from sin. In His Name. Amen.

REF.: Genetic key to perfect pitch. Minneapolis *Star Tribune*, Nov. 20, 1990.

The Hungry Leaf

Ephesians 2:10

"For we are his workmanship, created in Christ Jesus unto good works, which God hath before ordained that we should walk in them."

It's a warm day on the Amazon River. A small fish swims near the surface looking among the natural floating river debris for some food. As he looks among the occasional leaf or twig in the water he sees no danger. At the same time, one of the floating "leaves" is waiting for the little fish to move just a bit closer. Then, without warning, the floating "leaf" comes to life, grabs the fish and eats it!

What the little fish thought was simply another floating leaf was actually another fish. The Amazonian leaf fish is carefully designed not only to look like a leaf, but to act like one as well. It has a flat body, very much like a leaf. A black line runs the length of its body, giving the appearance of the midrib of a leaf. A fleshy growth in its lower jaw looks like a leaf stem.

Beyond looking like a leaf, the leaf fish also acts like a leaf. It lies still in the water, drifting with the current. To hide its identity further, it draws its fins close to its body, removing any last hint that it is indeed a hungry fish looking for lunch.

The leaf fish combines deceptive coloring, appearance and behavior into one design so that it, too, can make its living. In this the leaf fish shows that the Creator makes nothing without also designing a special purpose for that creature. According to the Bible, the same can be said even more emphatically for each human being He has made. Your Creator has a special purpose for you, too, and it begins with His plan of salvation for you through His Son, Jesus Christ.

Prayer: Dear Father in heaven, through the instruction of Your Word help me to better learn how to define who and what I am more completely in the knowledge of Your forgiving love to me in Jesus Christ. Amen.

REF.: Hanson, Jeanne K., and Deane Morrison. 1990. The wonderful weird of flora and fauna. *Star Tribune First Sunday*, Dec. 2. p. 17.

Multiple Blessings from Trees

Genesis 1:11

"And God said, Let the earth bring forth grass, the herb yielding seed, and the fruit tree yielding fruit after his kind, whose seed is in itself, upon the earth: and it was so."

Genesis uses only a few words and simple language to describe the creation of trees. That hardly seems fair, considering the importance, variety and complexity of the largest and longest-living things on earth.

The largest Sequoias in northern California weigh more than six blue whales. The tallest Australian eucalyptus trees and redwoods tower as high as a 30-story building. Some of the Bristlecone pines in the American west are 4,000 years old.

Trees do more than provide us with building materials, food and medicine. Like other green plants, trees take the waste carbon dioxide produced by animals and humans and turn it into oxygen. As a bonus, trees produce sugar. The cellulose fiber we call wood is actually only thousands of sugar molecules linked into long chains. In recent years scientists have also learned that trees communicate with each other. They cool the air and pump enough moisture into the air to transform a dry desert-like climate. If it were not for the *mycorrhiza* fungi, trees would battle through their roots for control of moisture and soil resources. However, when this fungus is present in the root systems, trees will link roots with each other and share water and nutrients.

Scripture's simple statement that God created trees expands into volumes of information as we learn more about them. This is how science done in faith—that is, as our search for knowledge about what God has made—serves to glorify our Creator!

Prayer: Dear Lord, while Your Word is simple enough to be understood by a child, it is also exact enough to impress the scientist. I ask You that Your Word would be accessible in this way to all. Amen.

REF: Ponte, Lowell. What good is a tree? *Reader's Digest.*

Scientist Says Evolution Makes Bad Science

Isaiah 22:11

"Ye made also a ditch between the two walls for the water of the old pool: But ye have not looked unto the maker thereof, neither had respect unto him that fashioned it long ago."

Since the days of Charles Darwin, over a century ago, naturalists have generally classified living things by their appearance. A zebra and a horse are said to be more closely related than a zebra and a whale. This has been the foundation of all those evolutionary trees that claim to show that all living things have supposedly come from older, simpler living things.

In recent years some scientists have been comparing another feature that all living things share. Today, scientists are able to look for similarities and differences among the biological chemicals like hemoglobin. What they are finding does not agree with the evolutionary tree in their textbooks, according to Dr. Colin Patterson of the British Museum of Natural History.

Among the examples Dr. Patterson offered to American evolutionary scientists was a comparison of alpha myoglobin among mammals, birds and reptiles. According to evolution, comparisons between mammals, reptiles and birds should show mammals and birds to be most distantly related. Reptiles should have more similarities to each. However, when humans were compared with crocodiles and birds, birds were closer to humans than were crocodiles, the opposite of what evolution predicts. Dr. Patterson asked, "What's going on?"

While Dr. Patterson has not become a creationist, he is dissatisfied with evolution.

Prayer: I thank You, Lord, that You have fashioned the creation so that it is difficult to deny Your work. As the creation speaks of Your handiwork, use my voice to witness to the forgiveness of sins in Your death and Resurrection. Amen.

REF.: Sunderland, Luther. 1988. *Darwin's Enigma*. El Cajon, CA: Master Books.

The Ant Imposter

Ephesians 4:14

"That we henceforth be no more children, tossed to and fro, and carried about with every wind of doctrine, by the sleight of men, and cunning craftiness, whereby they lie in wait to deceive;"

One good method of avoiding attack in nature is to pretend that you are somebody who should not be attacked. In the insect world, one of the most feared creatures is the ant. Not only are ants ferocious, they also taste bad.

As a result, many insects mimic ants to avoid predators. One of the most interesting groups of ant mimics is spiders. Of course, even insects can count, so they know that ants have six legs while spiders have eight. To solve this problem, some ant-mimicking spiders wave their front two legs over their heads so they look like antenna. The ants don't seem to be fooled, however. Nevertheless, they will allow the spiders to hang around so long as they stay out of trouble.

Some red ants have red spiders who mimic them and black ants have some black spiders who mimic them. This suggests that many insects can see color. Wasps that prey on one African red spider are no match for the vicious red ant the spider sometimes mimics. The African red ant can easily kill the wasp. This wasp also preys on a black spider. So the black spiders are often found among black ants that are almost as dangerous as the African red ants.

Some modern scientists tell us that this appearance of intelligence in insect behavior is nothing more than appearance. However, we should not be as easily fooled as these wasps are by the spiders. There are millions of similar evidences of intelligence in the creation, and they can't all be deceptive appearance!

Prayer: Dear Father, instruct me by Your Word and strengthen me by Your grace so that I may not be easily deceived by the devil, the world, and my own flesh. In Jesus' Name. Amen.

REF.: Hanson, Jeanne K., and Deane Morrison. 1990. The wonderful weird of flora and fauna. *Star Tribune First Sunday*, Dec. 2. p. 17.

Creatorless Education Rejected by Authors

Romans 13:2
"Whosoever therefore resisteth the power, resisteth the ordinance of God: and they that resist shall receive to themselves damnation."

Why can't Johnny read? Why can't he do simple math or find North America on a globe? In fact, why do Johnny and Jane reject education today? While parents and education officials in Washington are asking these questions, one of the authors of the current mess is taking responsibility for his mistake. He is also offering solutions.

Dr. William Coulson, working with Carl Rogers and Abraham Maslow, developed the basis of modern education, grounding it in the humanistic principles of John Dewey. Even if you don't recognize these names, you will recognize their ideas. For example, we have all heard the claim that the nondirective sex education courses in today's public schools can solve the teen pregnancy problem. Yet every study shows that these classes increase teen pregnancy. And we have all heard of values clarification.

Dr. Coulson has been spending the last three years telling all who will listen that the fundamental problem with today's educational philosophy is that it rejects authority. Once students are taught to reject the Creator's authority, they quickly reject all other authority, including their teachers. Dr. Coulson says that, as a result, students are not only unteachable, but they lose the ability to make moral choices.

Christian parents as well as educators desperately need to hear Dr. Coulson's message.

Prayer: Father, we only destroy ourselves when we reject You. Yet that is the nature of all sin. Create in me a clean heart and renew in me a right spirit. In Jesus' Name. Amen.

REF.: Pearcey, Nancy. 1991. The evolving child: John Dewey's impact on modern education. *Bible-Science Newsletter*, Jan. p. 6, Feb., p. 5.

Smart Trees and Clever Bugs

Psalm 104:24

"O LORD, how manifold are thy works! in wisdom hast thou made them all: the earth is full of thy riches."

How smart is the smartest tree you ever met? Does it sound silly to speak of trees as being smart? Botanists are learning that some trees have elaborate defenses against insect pests.

Studies have shown that a cottonwood tree sets up an elaborate zone defense when attacked by aphids. Some branches will manufacture poisons to resist the aphids while others use a different strategy. Branches that offer the aphids poisons can be dozens of times more resistant to aphids than neighboring branches. Should an aphid land on a leaf that is not manufacturing poisons, that leaf is very likely to be shed by the tree. This squeeze play by the tree keeps both strategies viable in resisting aphids.

However, sometimes insects are almost as clever as trees. The eastern tent caterpillar eats a fatal dose of cyanide with each mouthful of cherry tree leaves without harm. The poison is combined with two sugar molecules in the leaves. Normally an insect's digestive system would separate those sugars, releasing the cyanide. However, the tent caterpillar's digestive system releases only one sugar, leaving the cyanide bound and harmless. On the other hand, the pink sawfly larva concentrates and stores the toxic oils of the eucalyptus leaves it eats. When threatened, the worm-like larva rears up and spits droplets of the poisonous oils at its attackers.

There is no level at which the creation fails to show forth intelligence and planning. Every level of creation glorifies God!

Prayer: Father, forgive me for my failure to thank and praise You. I especially thank You for the gift of the forgiveness of my sins through the innocent suffering, death and resurrection of Your Son, Jesus Christ. Amen.

REF: Brownlee, Shannon. 1983. The silent battle. *Discover*, Sept. p. 28.

The Rock-Climbing Catfish

1 Peter 4:11

"If any man speak, let him speak as the oracles of God; if any man minister, let him do it as of the ability which God giveth: that God in all things may be glorified through Jesus Christ, to whom be praise and dominion for ever and ever. Amen."

It's not very pretty, nor is it graceful, but Colombia's rock-climbing catfish has a unique ability among fish. Perhaps there are days when a salmon, trying to make it up a waterfall, wishes that it could simply climb the rocks.

The rock-climbing catfish's climbing equipment begins with a powerful sucker mouth. Once it attaches itself to a rock, it can stay stuck despite any torrent. The rock-climbing catfish also has a bony plate on its belly that helps it hold its place while it repositions its sucker a little further up the rocks. The plate has fins with tiny, backward-pointing teeth. These teeth dig into the rock and prevent slipping. The catfish's third special feature is a set of powerful muscles to control the plate.

Once the rock-climbing catfish has attached its sucker to the rock, it pulls up its stomach plate and anchors it. With its body anchored, it moves its sucker up the face of the rock and repeats the process. Following this method, the rock-climbing catfish can walk its way up a 20-foot rock wall against the torrent of a waterfall in about 30 minutes!

Who gave the rock-climbing catfish that powerful sucker mouth? Who gave it such a clever bony stomach plate and the powerful muscles to control it? And who taught it to climb rocks? There are two opinions today. One opinion says that *no one* did it. The other says that it was the work of the Creator of worlds and stars and you and me. The Bible's claim of a Creator is the only one that makes sense!

Prayer: Dear Lord, it is from You as the creating Word that all ability comes. Help me to use the abilities You have given me to glorify You in word and deed so that others may know of the forgiveness of sins. Amen.

REF.: Hanson, Jeanne K., and Deane Morrison. 1990. The wonderful weird of flora and fauna. *Star Tribune First Sunday*, Dec. 2. p. 17.

An Accent on Language Ability

Genesis 11:1

"And the whole earth was of one language, and of one speech."

After suffering a stroke, a man in Baltimore suddenly began speaking with a Scandinavian accent. Though he had no experience with languages other than English, he suddenly sounded Nordic and appeared unfamiliar with English! In the weeks after the stroke his speech gradually returned to normal. After three and one-half months, he again spoke with no trace of accent.

This affliction is real. It's called "foreign-accent syndrome." It's a rare condition in which a brain malfunction causes speech to sound as if the speaker has a foreign accent. Americans have also reportedly developed apparent German, Welsh, Scottish, Irish and Italian accents.

Scientists say the fact that speech follows certain patterns in this way suggests something about the way the human brain is wired. A growing number of studies and experiences—such as "foreign-accent syndrome"—suggest that humans have been programmed with much more than just language ability. Some of the unique features of specific languages may actually be programmed into our brains.

The Bible tells us that everyone on earth spoke the same language until God confused the languages of rebellious people at Babel. "Foreign-accent syndrome" suggests that many of the detailed characteristics of language may have been originally wired into the brain. It also helps us see that language is more that just noises to which we have given meaning. Language is God's gift.

Prayer: Dear Father in Heaven, I thank You for the gift of language. Forgive me for the times I have misused this gift and help me to once again use language in witness and praise to You. In Jesus' Name. Amen.

REF: Rare disorder after stroke puts the accent on brain understanding. Minneapolis *Star-Tribune*; Oct. 17, 1990. p. 7A.

Microscopic Engineers

Psalm 119:133

"Order my steps by thy word: and let not any iniquity have dominion over me."

There are a number of species of bacteria that possess tiny chains of crystal particles called *magnetosomes*. Magnetosomes, as the name suggests, have iron in them to help the bacteria sense magnetic fields for navigation. They can also help the bacteria find food.

These bacteria live in salty, sulfurous waters and sediments. Unlike most bacteria that have magnetosomes, the three species that prefer sulfurous waters don't make their tiny compasses out of iron oxide. Rather, they combine iron and sulfur to make a crystal called *greigite.*

Now it would seem astonishing enough to find bacteria that build their own compasses for navigation, but scientists had even more surprises coming as they learned about these three new species. They discovered that each of the species makes its own unique crystalline form of greigite. The bacteria seem to be able to build the crystals, molecule by molecule, at the normal temperatures and other conditions at which life exists. Scientists would like to know how these bacteria do this. Modern science can only make crystals under searing hot temperatures and other conditions that would kill anything alive.

These amazing abilities were not designed and built by mutations and accidents. They show that our Creator provides for the needs of all His creatures—even bacteria—and that He does so with unlimited creativity.

Prayer: Dear Lord, I thank You for Your greatest provision for me in Your holy, innocent suffering and death in my place for my sin. The wisdom of the plan of salvation confounds the enemy and amazes angels! Amen.

REF: Crystal growers seek bacterial know-how . . . *Science News*, v. 138, 1990. p. 382.

Do You Know What You Believe?

Acts 17:11

"These were more noble than those in Thessalonica, in that they received the word with all readiness of mind, and searched the scriptures daily, whether those things were so."

Are you sure you know what you believe? Do you know *why* you believe what you do? New research suggests that people generally do not know how they came to believe what they believe. Nor do they know why they believe certain things. This research confirms the truth of a biblical lesson.

A number of studies by researchers from the University of Texas at Austin suggest that there are subtle but important influences on how we come to believe certain things are true and other things are false. They say the first thing that happens when we hear something is that we believe it. Then we compare it with things we already know. In the case of a patently silly statement like "pigs can fly," the believing stage and the comparison with reality all happen in a split second. So when we reject the statement as false, we never realize how we came to that conclusion.

Researchers also found that if we are distracted during the acceptance and comparison stages, we might not finish with them, and we will accept a statement as true that we would normally reject. Beliefs can be created when we just passively accept information as true.

We can apply these conclusions to our Christian faith. Very often as Christians we will accept statements that contradict the Bible because our attention was distracted by other things that were also said—things that we may accept because they are biblical. This shows the wisdom of the Christians at Berea who, we are told, compared *everything* they heard with what the Bible had to say.

Prayer: I thank You Lord for the wonderful mind You have given me. Help me to be more careful to compare each element of what I hear to what the Bible has to say so that I am not misled. Amen.

REF: Bower, Bruce. 1991. True believers. *Science News*, v. 139, Jan. 5. p. 14.

Keeping Males and Females in Balance

Genesis 1:27

"So God created man in his own image, in the image of God created he him; male and female created he them."

Have you ever wondered how, over thousands of years, the population balance between males and females remains roughly even?

Science has learned that there seem to be several mechanisms working together to keep things balanced, providing a fail-safe system. Many species have gene transfer systems that always produce a one-to-one ratio in the next generation.

However, what happens if, perhaps due to a natural disaster, the balance is lost between males and females in a population? Scientists have been testing a theory that when either sex is significantly underrepresented, its increased contribution to the next generation brings things back into balance. This amazing theory is supported by recent studies on the Atlantic silverside fish. Studies show that even when the population has three males for every female, the ratio returns to 50-50 within four to six generations!

According to evolution, males and females should not have developed at all. Evolutionary doctrine says that if males and females should develop, evolutionary forces will not favor sexual reproduction.

The Bible says that God not only created male and female, but that the creation is designed with great wisdom. What science is learning about how balance is maintained between the sexes contradicts evolution and supports the Bible.

Prayer: Dear Father, You have made all things well! Help me to use the gift of sex in a way that always glorifies You that my life may be a witness for my Lord and Savior, Jesus Christ. Amen.

REF: Nature keeps male-female ratio steady. *Chicago Sun-Times*, Dec. 1990. p. 56.

Salty Enough for the Bible

Colossians 4:6

"Let your speech be alway with grace, seasoned with salt, that ye may know how ye ought to answer every man."

How much salt should there be in the ocean? We know how much salt rivers carry into the sea each year. And we know how much salt is already in the ocean. So the question is, how long has salt been building up in the ocean?

This topic was the subject of an impressive paper presented at the Second International Conference on Creationism in 1990 in Pittsburgh. The research was presented by Dr. Steven A. Austin and Dr. D. Russell Humphreys, scientists who believe in creation. Even some of the evolutionists attending the conference admitted that the paper raised some challenges to evolutionary claims about the age of the oceans.

This new research carefully looked at all the sources that put salt into the ocean and all the processes that remove salt from the ocean. Evolutionists say that salt is being added to the oceans at about the same rate as it is being removed. If this were not the case, the oceans could not possibly be billions of years old. However, creation scientists Austin and Humphreys showed that salt is being added to the ocean and probably has been in the past. In fact, the oceans could not possibly be even 1/50 of the age claimed by evolutionists. Their conclusions also call radioactive dating methods into question.

We don't need to know any science to know that the truth will never contradict the Bible. However, we should also know that scientists who believe in creation can offer impressive scientific challenges to the claims of evolutionary science.

Prayer: Dear Lord, I thank You that You have given us some scientists who believe that Your Word is true in all that it says. I ask that Christians, including myself, would be made bold in our witness to Your Word. Amen.

REF: Austin, S.A., and D.R. Humphreys. 1990. The sea's missing salt: a dilemma for evolutionists in *Proceedings of the Second International Conference on Creationism*. Pittsburgh, PA: Creation Science Fellowship, Inc.

Unimagined Actions and Reactions

Job 10:12

"Thou hast granted me life and favour, and thy visitation hath preserved my spirit."

The study of plants' abilities to protect themselves from insect attackers is a fairly new and rapidly growing area of research. It seems that new and even more amazing discoveries are appearing every month.

Let's imagine that you are a nice little beetle looking for a tasty tomato plant. Soon after spotting a juicy looking tomato, you are happily munching away. However, before you can finish your meal, you are in trouble. Every leaf on the tomato has started turning out chemicals that will, at the least, give you serious indigestion and possibly kill you. At the same time, the plant has warned its neighbors of your presence, and the other tomatoes, too, are turning out the same nasty chemicals. Your very act of eating has made it impossible to get a decent meal in this garden!

The latest research suggests that the tomato under attack communicates with its neighbors by releasing a chemical called methyl jasmonate. (This chemical is commonly found in perfumes.) Methyl Jasmonate warns neighboring plants that hungry insects are in the neighborhood. On receiving the news, they begin producing the two chemicals that prevented you from digesting the protein in your lunch. This reaction has been seen among a number of plants, including cotton plants and various trees.

Who would ever have imagined that plants warn each other about attack? Who would have guessed that they know enough chemistry and insect biochemistry to mount such a clever defense? While even we who believe in creation are amazed, we know that the Creator can and did equip plants with these abilities.

Prayer: Dear Father in heaven, we know that life is more than just a chemical reaction. All life reflects Your care and wisdom, characteristics that cannot come from matter itself. I thank you that this is so evident. Amen.

REF: Chen, Ingfei. 1990. Plants bite back. *Science News*, v. 138, Dec. 22 & 29. p. 408.

Plants Call in Air Defenses

John 1:14

"And the Word was made flesh, and dwelt among us, (and we beheld his glory, the glory as of the only begotten of the Father,) full of grace and truth."

New research shows that, when attacked, many plants make chemicals that will give munching insects a bad case of indigestion. They also seem to warn their neighbors about the attack so that they, too, start up their defenses. Some plants even call in air defenses.

Some plants, when attacked by caterpillars, release a scent that scientists simply describe as "green leaf odors." These odors attract certain female wasps, who home in on the plant. The wasp will sting the caterpillar, leaving him paralyzed. She then lays her eggs in the pest, who remains paralyzed until the hatching young wasps consume it. That "green leaf odor" is a plant's way of calling in air defenses. It is communication between plant and insect!

Some evolutionary biologists are upset with the intelligence and design in this arrangement. Plants are supposed to be too simple to communicate. As one evolutionist put it, this is "not a defense against anything," it "just happens."

I remember trying that excuse as a child. The cookie jar fell to the floor while I was trying to sneak a cookie. "It just happened," I said, as if the cookie jar had come alive before my unbelieving eyes and jumped to the floor. My mother reminded me that things *don't* "just happen." And science teaches us that things *don't* "just happen." Not only do we see that God exists, we also see a bit of His love in providing for the plant and His intelligence as the author of communication.

Prayer: Dear Lord, as the eternal Word, You were the Instrument through which creation took place. When we ruined ourselves with sin, You took upon Yourself the human flesh You had created to save us. Thank You! Amen.

REF: Chen, I. 1990. Pest-eating allies: calling up the reserves. *Science News*, v. 138, Dec. 22 & 29. p. 410.

A Discovery Predicted by Creationists

Genesis 1:31-2:1

"And God saw every thing that he had made, and, behold, it was very good. And the evening and the morning were the sixth day. Thus the heavens and the earth were finished, and all the host of them."

The ability to predict new discoveries is an important test of whether someone is being scientific in their approach to knowledge. In science, it is especially desirable to predict correctly an unexpected discovery. Many theories that were rejected by almost all of science became accepted because they correctly predicted unexpected discoveries.

Not too long ago, evolutionists did not believe that any mammals lived at the same time as dinosaurs. Creationists knew that all the different kinds of living things appeared during the first week of earth's history. They have long been predicting that as more fossils are discovered, the age of so-called "modern" creatures would become older and older. They have even explained why, based on the Flood.

Over the years, evolutionists have had to admit that some mammals were alive while dinosaurs were still living. Now, a new discovery has pushed back the age of anthropoids by 50%. That places anthropoids fairly close to the age of the dinosaurs. According to evolution, anthropoids—like tarsiers—are ancestors of apes, monkeys and humans.

What is even more amazing is that the fossilized tarsier is said to be nearly identical to tarsiers today. This discovery offers strong support for the views held by scientists who believe that the Bible's account of history is true. And creationists continue to predict that the age of all living things will be pushed back toward the appearance of the first life.

Prayer: Dear Father, I thank You for those scientists who believe the truth of Your Word. Use their work and words, along with my words, to witness Your truth of salvation in Jesus Christ. Amen.

REF: Discovery of tiny primate fossil puts new spin on human evolution. St. Paul, MN, *Pioneer Press*, Jan. 4, 1991.

The All-Purpose Foot

Psalm 44:8

"In God we boast all the day long, and praise thy name for ever. Selah."

Nothing is more frustrating than glue that refuses to stick. And nothing is harder on an adhesive's ability to stick than water. Water can prevent adhesives from solidly touching the surfaces that are to adhere. Freezing and thawing temperatures can force glued surfaces apart. Yet biological adhesives constantly work without fail in continual contact with water.

The mussel is the champion in the creation and use of biological adhesives. When a mussel finds a suitable spot, it extends its all-purpose foot. The foot, which is like a rubber plunger, searches the surface until it finds the exact spot that will make the best bond. The foot then cleans the glue point. Next, the foot presses to the surface and as the plunger-shaped foot flattens, water is pumped out. The mussel then lifts the center of its "plunger" and creates a small vacuum. Now the ingredients for its adhesive are pumped down through the foot and into the vacuum. This material creates a foamy foundation for the individual glue threads.

The glue itself is made of several proteins. They are mixed in various ratios to provide just the right amount of strength, flexibility and resistance to compression needed for each unique anchoring spot. Some of these proteins are very unusual in structure. Scientists suspect that they change their properties in response to changing conditions—what scientists call a "smart" material.

The mussel is not only a wonder of design, but a sophisticated chemist as well. Modern science is still trying to learn what the Creator has taught the mussel about glue.

Prayer: Dear Lord, when science tries to learn how the mussel makes its glue, it is recognizing that the creation has sophisticated designs in it. Even if it's unintentional, this is praise for You. I praise You, too. Amen.

REF: Amato, Ivan. 1991. Stuck on mussels. *Science News*, v. 139, Jan. 5. p. 8.

Bacterial Physicists and Chemists

Exodus 3:6

"Moreover he said, I am the God of thy father, the God of Abraham, the God of Isaac, and the God of Jacob. And Moses hid his face; for he was afraid to look upon God."

Some types of bacteria are able to go into a state of suspended animation when living conditions become too harsh. Not only do they change their form and appearance, they are amazingly able to change their DNA structure to avoid damage from harmful types of light!

Like other living things, bacteria need the right temperatures, the right amount of moisture, and other conditions that support life. When conditions change, bacteria are too tiny to move to better conditions. However, they can *sporulate*. When living conditions become too harsh, some bacteria encase themselves into a tough coating of protein and enter a state of suspended animation. As spores, bacteria can rest in bone dry soil for up to 60 years waiting for rain.

Ultraviolet radiation poses another danger to bacteria. This light damages a bacterium's DNA at a point where two thymine molecules sit next to each other. The light strikes the molecules and twists them. The result can be fatal to bacteria. So when a bacterium begins to go into spore form, it produces a protein that slightly and harmlessly changes the shape of these two thymine molecules so that they are not affected by ultraviolet light.

The entire process by which bacteria go into spore form reflects a sophisticated knowledge of both biochemistry and physics. Even in the lowly bacteria, our Creator has made it clear that He and He alone deserves the credit for the creation in which we live!

Prayer: Dear Lord, the astonishing care with which You made even bacteria assures me that there is nothing in Your creation that You do not care about. Assure me constantly of Your presence and love in my life. Amen.

REF: Weiss, R. 1991. New light on bacterial DNA protection. *Science News*, v. 139, Jan. 12. p. 21.

Living Stones and Flying Flowers

Colossians 2:4-5

"And this I say, lest any man should beguile you with enticing words. For though I be absent in the flesh, yet am I with you in the spirit, joying and beholding your order, and the steadfastness of your faith in Christ."

If, of all the millions of living things on earth, a *few* creatures happened to look like something they are not and this helped protect them, we might put it down to chance. However, the creation is filled with too many creatures who practice look-alike deception to put it down to chance.

Africa's southern deserts have a plant called living stones. In a desert, any vegetation can be a welcome source of food or moisture. In this setting, a plant that doesn't look like a plant has an advantage. And living stones look like nothing more than just another rock in the sand. Wasps can be some of the nastiest insects in nature. An insect in Costa Rica called the mantispid looks very much like the feared wasp, even up close. Then again, the hover fly looks like a wasp and sounds like a wasp, because the frequency of its wing beats is almost identical to a wasp.

Treehoppers have a very odd body shape that comes to a point on top. Not only do they look like thorns, but they usually stay on bushes that have thorns, orienting their bodies on the branches just like true thorns. This effectively hides them from hungry birds. Larval plant hoppers use a different form of deception. They attach themselves in groups on branches, orienting their frilly and feathery bodies together so they look like a flower!

The clever forethought and planning involved in nature's deceptions rule out any thought that these creatures designed themselves.

Prayer: Lord, I know that the devil knows a great deal about deception, too. I confess that I have been fooled into sin that did not look like sin. Forgive me and make me wiser by Your Word. Amen.

REF.: Four masters of deceit. *National Geographic*, Mar. 1980. p. 410.

Your Busy Brain

Psalm 25:4

"Shew me thy ways, O LORD; teach me thy paths."

Not only is the human brain the most complex structure in the universe, it is also one of the busiest. In fact, during intense concentration, your brain can burn as many calories as your muscles do during exercise! Intense thinking can literally be as exhausting as a physical workout! More than 100,000 chemical reactions go on in your brain every second.

Among the brain's many jobs is chemist. The brain produces more than 50 psychoactive drugs. Some of these are associated with memory, others with intelligence, still others are sedatives. Endorphin is the brain's painkiller, and it's three times more potent than morphine. Serotonin is produced by the brain to help keep our moods under control. The brain also makes dopamine. Dopamine makes people more talkative and excitable. Another hormone regulates hunger.

The brain is also a radio transmitter that sends out measurable electrical wave signals. In fact, the brain continues to send these signals for as long as 37 hours after death!

Our tiny knowledge of the brain is enough to show us that this incredible organ is no accident. The brain is much more than a powerful testimony to our Creator. Its incredible powers convince us that there is much more to existence than this material life. However, we can learn only from the Word of God about eternal life through the forgiveness of sins in Jesus Christ.

Put your brain together with the Word of God today!

Prayer: Dear Father, I thank You for the incredible gift of my brain. Forgive me, for Jesus' sake, for my misuse of this gift and help me to better use it in Your service. In Jesus' Name. Amen.

REF.: McCutcheon, Marc. 1989. *The Compass in Your Nose...* Los Angeles: Jeremy P. Tarcher, Inc.

Witchweed

Matthew 11:28

"Come unto me, all ye that labour and are heavy laden, and I will give you rest."

The strange and nasty plant called witchweed is found mainly in Africa and Asia, but some has been reported in North America as well. This parasitic plant can prevent crop plants from producing yields and is said to be responsible for intensifying the results of famine in Ethiopia.

Not long after germination, witchweed sends an enzyme "feeler" through the soil around it to find nearby roots. If the enzyme finds a root system, it will digest a few of the cells on the root surface. The damaged roots actually help the witchweed by sending out a chemical response that confirms their presence to the witchweed. The witchweed then grows an attachment organ to penetrate the neighbor's root system. However, witchweed cannot penetrate many root systems. If attachment is not made within four days, the witchweed grows aimlessly and dies in less than a week.

If the witchweed successfully attaches to a neighbor's root system, witchweed can enjoy a long and full life being fed by its neighbor's roots. In fact, witchweed changes the hormones produced by its neighbor so that the neighbor's root system grows ever larger at the expense of the stem growth of the victim. While the victim may live its normal span, it will not be allowed enough strength to reproduce.

Witchweed is no simple plant. We learn from the witchweed that the Creator's unlimited imagination not only created some very strange living things, but also created some unique solutions to providing a means of living for them.

Prayer: Dear Father in heaven, I face problems in my life. I commend all of my cares and concerns into Your powerful, inventive and loving care. In Jesus' Name. Amen.

REF: The underground sorcery of witchweed. *Discover*, Dec. 1985. p. 14.

A Master of Disguise

Romans 1:18-19

"For the wrath of God is revealed from heaven against all ungodliness and unrighteousness of men, who hold the truth in unrighteousness; Because that which may be known of God is manifest in them; for God hath shewed it to them."

Many plants and animals disguise themselves as something they are not in the hope of protecting themselves. However, a true master of disguise must have many disguises. The tiger swallowtail offers no less than three disguises as it grows through its various stages toward adulthood.

The newly hatched larva looks for all the world like a bird dropping. This appearance makes the larva of no interest to birds that might otherwise eat it. Three molts later, the caterpillar has turned green to match the leaves on which it feeds. In case he is spotted by some birds looking for a nice fat caterpillar for lunch, the caterpillar also has two large spots on its head that look like the eyes of a snake. Birds who are interested in caterpillars usually try to stay as far from snakes as possible. Finally, in the pupal stage, the tiger swallowtail has again radically altered its appearance. The pupa now looks like nothing more than just another broken twig on a tree trunk.

The immature tiger swallowtail's bag of disguises reflects a great deal of knowledge about the behavior of those creatures that threaten it. Each of its three disguises uses a different method to fool those birds that would be interested in the larva. Each disguise has the same specific goal. This master of disguise is no accident of biology.

Our Creator's purpose in making Himself evident in the creation is ultimately aimed at bringing each of us into a relationship with Himself. That relationship is based on the forgiveness of our sin through His Son, Jesus Christ.

Prayer: I thank You, Lord, that You seek us so that we may not remain in our sin. Assure me of Your forgiving love and grant me Your peace through the forgiveness of sins. Amen.

REF: A one-bug virtuoso starring in three roles. *National Geographic*, Mar. 1980. p. 402.

Insulated Ears

Luke 11:28

"But he said, Yea rather, blessed are they that hear the word of God, and keep it."

Our senses link us to the world around us and enable us to interact with the world. While most people believe that our sense of sight is the highest of the senses and the most marvelous in design, our sense of hearing is no less marvelous.

When a sound strikes your ear, your eardrum vibrates with the sound waves, fast or slow, soft or hard. These variations in vibration provide us with important information about the nature of the sound we are hearing. Some sounds produce a vibration in the eardrum as small as a billionth of a centimeter—only one-tenth the diameter of a hydrogen atom! There are three tiny bones in the middle ear called the hammer, anvil and stirrup. They pick up the vibrations from the eardrum, amplify them and send them on to the cochlea. The cochlea is filled with about 25,000 tiny hair cells that finally turn the vibrations into electrical signals that are sent on to the brain.

Our hearing is designed to be more sensitive to high sounds than to lower sounds. If we had just a little more sensitivity to lower-pitched sounds, we would continuously be distracted by the internal sounds of our body, including the blood rushing through our arteries. In fact, to help prevent this, there are no blood vessels at all in that part of the ear where vibrations are turned into electrical impulses. The body supports life in these tissues by constantly bathing them in dissolved nutrients.

While we can hear many things, and decide not to hear other things, the Creator who so marvelously designed our ears invites us to hear His Word to us from the Bible.

Prayer: Father, I thank You for Your Word, which tells me of Your love for me in Jesus Christ and Your will for my life. Forgive me for those times I have chosen not to hear Your Word, and let me be more eager to hear it in the future. Amen.

REF.: McCutcheon, Marc. 1989. *The Compass in Your Nose...* Los Angeles: Jeremy P. Tarcher, Inc. p. 90.

Heated Moths

Matthew 6:30

"Wherefore, if God so clothe the grass of the field, which to day is, and to morrow is cast into the oven, shall he not much more clothe you, O ye of little faith?"

We tend to associate being warm-blooded with mammals and human beings. However, past "Creation Moments" programs have dealt with "warm-blooded" plants, reptiles and bumblebees. There is also a warm-blooded moth.

Since plants don't have blood, you might suspect that we are here using a broader meaning for the term "warm-blooded." When we use the term "warm-blooded" in referring to plants and insects, we are talking about a plant or animal that can generate enough body heat to carry on life normally, even if the air temperature is below freezing. There are about 50 species of moths in North America that are active in the cold of winter.

It takes a good deal of energy to generate heat for such a tiny body, since the heat is rapidly lost in the cold air. Most moths feed on nectar that is not available in the winter. The mystery deepens with the fact that the moth could not possibly store enough energy in its body to generate heat. Naturalists now have strong suspicions that these "warm-blooded" moths are involved in one of the strangest interrelationships in nature. They believe that the moths get their energy from sugar maple sap that is tapped and made available by red squirrels!

How the red squirrels do this is the topic of the next "Creation Moments" program. For now, these warm-blooded moths teach us that the features we normally associate with more advanced creatures actually provide no evidence whatsoever for evolution.

Prayer: Dear Lord, I thank and praise You for the wonderfully creative ways in which You provide for all of Your creatures. Help me to remember Your care when I am concerned about the needs of my material life. Amen.

REF: Heinrich, Brendt. 1991. Nutcracker sweets. *Natural History*, Feb. p. 8.

Smart Squirrels

Matthew 7:11

"If ye then, being evil, know how to give good gifts unto your children, how much more shall your Father which is in heaven give good things to them that ask him?"

Squirrels often seem to run about without any purpose. However, a naturalist in the State of Maine recently discovered that the squirrel's activities are well planned and efficient.

It was a cold January day when the naturalist was watching a red squirrel who seemed to be nibbling on a sugar maple. Then the squirrel would run to a tree where some of the sap had leaked out and lap at it with his tongue. The naturalist knew that conventional wisdom said it was too early to tap trees for sugar. He also knew that the sap from the trees is 98 percent water and could not provide enough energy to keep the squirrel going in cold weather. Becoming more curious, he decided to study the squirrels' activity in detail.

He discovered that the red squirrels were systematically tapping the sugar maples with a peculiar bite into the tree. Each bite went through the outer and inner bark and into the xylem where the sap runs. The squirrels were then allowing the running sap to dry until the sugar content was more than 55 percent. Not only were squirrels tapping just the maples, they only tapped them when the weather was right for the sap to run. The mad running up and down the trees actually followed a careful daily schedule of tree visitations that could include over 70 trees!

Could it be that humans learned from squirrels to tap sugar maples? Whatever the answer to that question, we know that it was the Creator Himself Who taught the squirrel the processing and use of this sugary winter treat.

Prayer: Dear Father in heaven, I thank You that You not only give us things to make life more pleasant, but that You are equally generous with all Your creatures. Let me never be ungrateful to You. In Jesus' Name. Amen.

REF: Heinrich, Brendt. 1991. Nutcracker sweets. *Natural History*, Feb. p. 4.

Inchworm Games

Isaiah 40:8

"The grass witherth, the flower fadeth: but the word of our God shall stand for ever."

We have looked at many plants and animals that use different forms of camouflage to keep from being noticed. Some don't wish to be noticed so that they can surprise their prey, while others don't want to be noticed so they can avoid becoming prey. However, what happens when both predator and prey encounter each other in full camouflage?

The predator in this drama, witnessed by a naturalist at the edge of a desert, is a yellow crab spider. The spider was waiting on a yellow flower that made him nearly invisible. The potential prey is an inchworm beneath the same flower. The inchworm was tearing off bits of the yellow flower petals and sticking them on its back.

Properly camouflaged, the inchworm then worked his way toward the top of the flower where the spider waited. The spider felt the movement and ran around the edge to see whether lunch might be on the way. However, when the spider got around the edge of the flower, he saw nothing. He didn't even notice the camouflaged inchworm on which he was standing! The spider then returned to the center of the flower, outsmarted by the inchworm who went on about his business.

Our wisdom would have said that the inchworm should try to run when he noticed the spider. However, the wisdom given to him by the Creator allowed him to continue about his business on the flower. This offers us a good illustration of how following the Lord's wisdom often requires us to reject our own.

Prayer: Dear Father, I know with my mind that You are far wiser than any human being. Forgive me for those times when I have applied my own wisdom to my life instead of Yours. In Jesus' Name. Amen.

REF: A predator's parlor games. *National Geographic*, Mar. 1980. p. 396.

Your Multi-Purpose Nose

Psalm 111:2

"The works of the LORD are great, sought out of all them that have pleasure therein."

Which body structure is a powerful generator of emotions, gives you a sense of direction, and serves as a cannon all at the same time? It's your nose, of course! It is common knowledge that without our noses the scent of a rose could not give us joy, nor would our food taste like very much. Less well known is the fact that our noses are directly hooked into the brain's limbic system. Your limbic system plays a central role in generating and controlling emotions. This is why certain scents are able to raise strong emotional memories in us.

The nose is also designed to filter out and catch particles that should not get into the lungs. When a particle is taken in through your nose, it is snagged by the tissue inside your nasal passages. Next, this highly special tissue decides the best method of getting rid of the foreign material. If a bacterium is identified, your nose begins to run, bathing the bacteria in powerful chemicals that dissolve most of them. If the particles are larger, like pollen, a sneeze is triggered that can eject particles at speeds of over 100 miles per hour!

Perhaps the most remarkable ability associated with the nose is its ability to act as a compass. Human beings have a small amount of iron in the bony part of the nose between the eyes. Scientific experiments have shown that we have an ability, while not foolproof, that can sense direction that is influenced by magnetic fields!

The Creator made your multi-purpose nose not only to bring some pleasure to life but also to help protect your health.

Prayer: Father, I thank You that just as You enrich our earthly lives with beauty and sweet scents, so we are also enriched in this life by the gifts of Your forgiving grace to us through our Lord Jesus Christ. Amen.

REF.: McCutcheon, Marc. 1989. *The Compass in Your Nose . . .* Los Angeles: Jeremy P. Tarcher, Inc. p. 95.

Designed Language

Genesis 3:9
"And the LORD God called unto Adam, and said unto him, Where art thou?"

Language requires some very special designs that are rare in nature and found together only in humans. First, human language requires vocal cords, which none of the apes have. The only animals with vocal tracts are birds, but no one is suggesting that we evolved from birds.

Language also seems to require an unusual structure to the brain. The frontal area of the left hemisphere of the brain is connected to an adjacent area of the cortex called the Wernicke area. These two areas are linked by a bundle of nerve fibers. The result is a distinct brain asymmetry that is unique to humans. Even the unborn human has this asymmetry that allows it to absorb language. Development in this area of the brain stops at puberty. That is why it becomes more difficult for adults to learn a new language. Language skills can be learned only by hearing language. So-called "feral" children, who have heard no language at all, cannot be taught language after puberty for the same reason.

So where did Adam first hear language? Scripture tells us that it was the Creator God Himself who regularly spoke with Adam. This means that language is not a collection of meaningless sounds to which we have assigned meanings. The meanings and sounds have come from God, just as our unique design that makes language possible comes from God.

Therefore language is not only for worldly purposes. We have also been given language to praise our God and Savior and to tell others who He is and what He has done!

Prayer: Dear Heavenly Father, forgive me for those times when I have misused language and those times when I have not used language to praise You. Let my words and worship be acceptable in Your sight for Jesus' sake. Amen.

REF: Tassot, D. 1990. Language-origins and evolution. *Creation Social Science and Humanities Quarterly*, Winter. p. 27.

Darwin Was No Geologist

Psalm 46:2

"Therefore will not we fear, though the earth be removed, and though the mountains be carried into the midst of the sea."

Charles Darwin's ignorance of geology, a science he never studied, probably resulted in the biological errors he made in formulating his theory of biological evolution.

As he sailed on the *Beagle* to the Pacific, Charles Darwin read Charles Lyell's book, *Principles of Geology*. That book, of course, theorized that the Earth's geology was the result of the slow processes we see today working over millions of years. On the way to the Galapagos he had a 16-day stopover in Argentina. He spent some of the time exploring the valley of the lower Santa Cruz River. He later wondered in his journal how the small and lazy Santa Cruz River could have carved the 300-foot-deep valley. But he allowed that Lyell's idea of long ages could solve that problem.

Later, at the Galapagos, he tried to explain plant and animal diversity based on those same long ages. Today, geologists believe that melting glaciers at the headwaters of the river formed a huge lake behind a natural dam. When that dam broke, the rushing lake water quickly cut a spillway that became the valley through which the Santa Cruz River gently flows today.

It appears as though Darwin's assumption that Lyell knew what he was talking about when he wrote of long geological ages influenced his biological interpretations.

Prayer: Lord, when I am intimidated by evolutionary challenges to the truth of Your Word, comfort me with Your promises. Amen.

Ref: *Acts & Facts*, 7/08, pp. 10-12, Steven A. Austin, "Red Rock Pass: Spillway of the Bonneville Flood."

Thomas Jefferson and the Creator

Psalm 100:3

"Know ye that the LORD he is God: it is he that hath made us, and not we ourselves; we are his people, and the sheep of his pasture."

Respect for the name Thomas Jefferson reaches far beyond the United States, and far beyond the English language.

By the time Jefferson was 14 he could read four languages. He later learned three more. Jefferson's legacy goes far beyond his accomplishments in law and government. He also made his mark in the fields of mathematics, architecture, medicine, agriculture, literature, education, music and science. He earned three doctorates. And he served as president of the American Philosophical Society, America's leading scientific organization of the time, for nearly 20 years.

One might think that Jefferson would have accepted the evolutionary ideas that were being circulated then by such people as Charles Darwin's grandfather. However, while Jefferson was a deist rather than a Christian, he spoke and wrote *against* evolutionary ideas and in favor of a Creator. He based his argument that belief in a Creator is more scientific than evolution on the precise and interrelated design that is visible everywhere in the creation. If you would like more information, we recommend the book *The Real Thomas Jefferson*, published by the National Center for Constitutional Studies in 1981.

We thank God that His handiwork is so clear in the creation that one does not even have to be a Christian to know that He is Creator. His most earnest desire for all is that, seeing this much, we might be drawn to His forgiveness in Jesus Christ.

Prayer: Dear Father, grant me fresh and clear eyes to see with new wonder, every day, Your magnificent handiwork in the creation. Help me to thank You each day for Your forgiving love to me through Jesus Christ. Amen.

REF: Wehling, R.J. 1991. Thomas Jefferson speaks out on origins. *Bible-Science Newsletter*, v. 29:3, Mar. p. 1.

The Spider and the Fly

Exodus 15:11

"Who is like unto thee, O LORD, among the gods? who is like thee, glorious in holiness, fearful in praises, doing wonders?"

One would never expect to find any insect sharing living arrangements with a spider. That's why scientists were amazed to find a tiny fly in the rain forests of Central America who shares both room and board with a spider!

The fly spends most of his time on the back of the golden web spider, who does not seem to object. This spider does not deal with its victims in the same way most spiders do. When a large insect gets caught in the tough web of the spider, the golden web spider injects the victim with digestive juices to prepare it for eating. The fly keeps track of meal preparations from the spider's back. Just before the meal is liquefied to the spider's satisfaction, the fly buzzes over and drinks its fill in a few seconds. By the time the spider approaches lunch the fly is back napping on the spider.

Both the spider and the fly are designed with special features to make this most unlikely arrangement work. If the golden web spider's eating habits were like most spiders', there would be no liquefied meal for the fly. The fly itself has a mouth that is specially designed to allow it to drink a full meal in seconds. This helps keep him out of the spider's way.

How can a spider and a fly live together? God can design a solution to any problem. Why design an arrangement between a spider and a fly anyway? So that we would know there is a Creator and that He can meet any challenge! There is nothing you cannot take to Him in the name of His Son, our Lord and Savior, Jesus Christ.

Prayer: Dear Father in heaven, I thank You for Your marvelous work of creation. But even more marvelous and a greater cause of thanksgiving for me is Your forgiving love to me through Your Son, Jesus Christ. Amen.

REF: Random chance or intelligent design? *Associates for Biblical Research Newsletter*, Jan.-Feb. 1991. p. 2.

The Jellyfish with a Broad Reach

Genesis 1:21

"And God created great whales, and every living creature that moveth, which the waters brought forth abundantly, after their kind, and every winged fowl after his kind: and God saw that it was good."

Do you know how to handle a sailboat? If you had to sail to a destination that was not exactly downwind, would you be able to do it?

Unlike most jellyfish, *velella* does not merely drift with the wind. *Velella* has a triangular sail. However, unlike some jellyfish with sails, *velella*'s sail doesn't simply catch the wind and move *velella* along with it. *Velella*'s triangular sail has an S-shaped base so that it can stand up to winds from any direction. The sail is set on top of *velella*'s body at an angle to its length. This makes *velella* a biological oddity, because its body is not symmetrical.

This asymmetry is not without purpose. The unusual body design of this jellyfish allows it to sail as much as 60° off the direction of the wind. This is what sailors call a broad reach. Some *velellas* have their sails tilted to the right, allowing them to sail up to 60° to the right of downwind. Others have their sails tilted to the left, allowing them to sail up to 60° left of downwind. What's more, *velellas* can change their angle of sailing by extending or retracting their tentacles in the water.

When scientists noticed that *velella*'s body was asymmetrical, they immediately looked for a purpose behind the design. Even though many scientists will not recognize the Creator, they understand enough to realize that nothing in the creation is without purpose or design. So while many deny the Creator, they still recognize the pattern of His handiwork!

Prayer: Dear Lord, even though some reject You, despite their great knowledge of the creation, they cannot deny the principles of purpose and design with which You have made all things. For this I glorify You. In Jesus' Name. Amen.

REF: The by-the-wind sailor: a masterpiece of natural design. *Discover*, Aug. 1987. p. 46.

The Impossible Gene

Acts 20:32

"And now, brethren, I commend you to God, and to the word of his grace, which is able to build you up, and to give you an inheritance among all them which are sanctified."

Biologists have recently discovered a gene that suggests that God may not be above playing a practical joke on those who reject what His Word says.

Those scientists who believe that natural forces made everything, including you and me, think that we have inherited all that we are. However, the Bible says that all living things reproduce "after their kind." Thousands of years of experience in selective breeding have shown that plants and animals always reproduce after their own kind. Not only is every known example of a mutation harmful, but living things resist attempts to change them too much.

Geneticists have discovered a genetic basis for the Bible's statement that all creatures reproduce after their own kinds. Keep in mind that "species" and "kinds" are not necessarily the same. Fruit fly studies show that when some species of fruit fly try to mate with other species, the resulting offspring are either sterile females or die before maturity. In one instance, however, some males survive to adulthood. The so-called "rescue" gene that allows this to happen weakens the general health of the males. Carriers of the gene are sterile. This leaves a huge mystery for evolutionists. Think about it. Parents that carry this rescue gene *don't benefit* from it. Offspring that inherit it *can't pass it on.* As one evolutionist said, how could this possibly have evolved?

Living things do reproduce after their kind, just as the Bible says. In this instance, God has found a way to do this that challenges the very basis of evolution.

Prayer: Dear Lord, I thank You that my biological inheritance as well as my spiritual inheritance through the forgiveness of sins in Jesus Christ is from You and that these have not been left up to chance. Amen.

REF: On the rescue gene and the origin of species. *Discover*, Aug. 1987. p. 7.

Creation Makes Better Science

Psalm 111:10

"The fear of the LORD is the beginning of wisdom: a good understanding have all they that do his commandments: his praise endureth for ever."

Many people have been taught that evolution is science and creation is religion. However, the origins debate is really a debate between two religions. According to the evolutionist's own rules of science, evolution is not good science. Consider the claims of evolution in view of what science knows today.

Life has never been seen to develop from nonliving materials. Yet evolutionists say it did. Mutations, said by evolutionists to have created all the kinds of living things, have never been seen to produce one creature who was more complex or better able to survive. The moon has moonquakes, a magnetic field and internal heat. These are all indications that it is far younger than evolutionists care to believe. Geological strata were at one time sediments in water—flood waters. Evolutionists believe that the multiple layers of sedimentary rock were laid down one after another by local floods with millions of years between each flood. If that is true, how is it that these strata are often found showing the lower strata blending into the strata above it?

If language evolved, why are the most ancient languages the most complex? If religions evolved, why do the most ancient forms of religion worship one God—a Creator—while later forms of the same religions have many gods; gods who are much more like human beings?

Many more examples show that only the Bible's account of history makes sense in view of what we know.

Prayer: Dear Lord, the very stones cry out that You are indeed Creator and Almighty God. Help me to add my voice to this witness and tell others of Your forgiving love in Jesus Christ. In His Name. Amen.

Inchworm Strategy

Job 38:36

"Who hath put wisdom in the inward parts? or who hath given understanding to the heart?"

Some creatures simply try to look uninteresting to hungry predators. Others take their role playing so seriously that they actually fulfill some of the same roles as their models.

Loopers, or inchworms, move along the ground, twigs or branches with the characteristic opening and closing of an upside down letter *U*. These creatures are not true worms. Rather they are caterpillars. As caterpillars, they often look very much like the twigs or branches on which they move. However, their odd form of movement can give them away.

Here's an experiment you can try. If you ever notice a looper on a branch, simply give the branch a gentle tap. Chances are, the looper will suddenly freeze into the same shape as a nearby twig. If you were to carefully touch the looper, you would find that it had become as stiff as a twig. A botanist reports that while studying one looper moving along a branch, an ant happened along. Ants, of course, will eat loopers. The looper froze in a horizontal position between branches. The ant walked across the looper never noticing that its walkway was actually a potential meal.

Many scientists suggest that the looper "just happens" to be the same color as the branches on which it crawls. We ask, does it also accidentally know how to act like a twig to fool predators? The almighty hand of God is evident even in the inchworm.

Prayer: Dear Lord, all knowledge and wisdom comes from You. Grant me Your Holy Spirit so that through Your Word I may be taught more by You than I have allowed myself to learn in the past. In Jesus' Name. Amen.

REF: The looper—inching along from one disguise to another. *National Geographic*, Mar. 1980. p. 398.

Catching Bigger, Stronger and Faster Prey

Luke 1:37
"For with God nothing shall be impossible."

At first glance, jellyfish may not look like much of a creation. Many are quite beautiful. However, most jellyfish simply look like blobs of jelly, so fragile that what might be a minor blow to one of us could prove fatal to them.

There are around 500 species of jellyfish. They live everywhere in the oceans from the surface down to great depths. Some are as small as a BB. The largest jellyfish is the lion's mane, which can grow to seven feet in diameter and have tentacles 200 feet long.

Despite their apparent simplicity, these insubstantial creatures make their living by capturing and eating fish and crustaceans that could easily tear them apart. To accomplish this seemingly impossible task, the jellyfish trails out its long thread-like tentacles. Each tentacle can have millions of capsules loaded with tiny harpoons. Once touched, the tentacles fire, injecting several deadly toxins. If enough toxin is injected, the victim's heart stops. If not enough of this poison is injected, another toxin finds its way to the brain where it stops respiration. Yet another toxin bursts red blood cells, disrupting the circulatory system. Thankfully only a few dozen of the hundreds of species of jellyfish can hurt human beings.

Imagine trying to design a creature that is not much more than a bag of Jell-O that can capture and subdue much larger and more powerful creatures. The jellyfish may look simple. However, its specialized design for a seemingly impossible task makes it an elegant witness to our Creator.

Prayer: Dear Father in heaven, help me to remember that nothing is beyond Your ability and wisdom in the material world, especially when I face problems and am tempted to rely on myself for a solution. In Jesus' Name. Amen.

REF: Brownlee, Shannon. 1987. Jellyfish aren't out to get us. *Discover*, Aug. p. 47.

A Thousand Words for a Bee

Deuteronomy 6:5

"And thou shalt love the LORD thy God with all thine heart, and with all thy soul, and with all thy might."

They say that a picture is worth a thousand words. While that is often true, they also say that it takes a highly evolved brain to picture objects and remember them. Science is learning that this bit of evolutionary myth is indeed fiction.

As far as the honeybee is concerned, not all flowers are created equal. Some flowers are a good source of nectar while others are a waste of time. Alfalfa flowers can kill a honeybee, because they are designed for pollination by larger insects. When an insect lands, a central petal in the flower releases the stamen and sweeps pollen upward to aid pollination. Honeybees can be tossed off the flower or trapped inside by this action.

Scientists always thought that honeybees solved the problem of finding safe and productive flowers by remembering a list of characteristics. They believed the honeybee was evolutionarily too simple to remember mental pictures of favored flowers. However, recent experiments have convinced scientists that bees actually store a mental picture of productive flowers. In these experiments bees were rewarded with sugar water for visiting some patterns and not others. Their reactions showed that they remember images, not simply patterns.

What's more, the experiments showed that bees are quite intelligent and good learners. One scientist noted that these experiments showed that evolution's assumptions about so-called higher and lower creatures are false. Those who believe in evolution are having to rethink their view of the world. Meanwhile, we can give thanks to God for so generously giving the gift of intelligence to so many of His creatures.

Prayer: Lord, I thank You for the intelligence You have given me, and I ask that You would forgive me for the times when I have ignored and misused this great gift. Let me serve You with all my heart and soul and mind. Amen.

REF: To honeybees, a picture is worth a thousand line angles. *Science News*, v. 127. p. 196.

Pygmy Dates?

John 10:35b
". . . and the scriptures cannot be broken;"

Creationists are often objects of scorn because they take the Bible's account of history literally. If the Bible's history is accurate, humans were on the earth at the same time as many animals that are extinct today. Creationist arguments are often dismissed without answer by skeptics. Often they do this only because creationists believe that the entire creation is only about 6,000 years old. At the same time, we are gaining a more complete picture of the early earth's history. As a result, evolutionists are being forced to accept important parts of what creationists have always been saying.

Excavations in Cyprus have shown that humans have a long history of settlement there. Excavations have also revealed that Cyprus was once home to the extinct dwarf elephant and the pygmy hippopotamus. However, evolutionary scientists said dwarf elephants and the pygmy hippopotamus were there 1.5 million years before humans ever arrived on the scene. Creationists, of course, rejected this claim.

But in recent excavations, bones of pygmy hippos have been found in the same layers as chipped bone artifacts, chisels, scrapers, beads and hearths. Based on this evidence, project director Alan Simmons says, "I believe the earliest occupants of Cyprus enjoyed pygmy hippo steaks."

We may never know for sure whether the settlers on Cyprus ever really ate pygmy hippo steaks. However, evolutionists now have to stop laughing at creationists long enough to admit that at least this part of our young earth position is scientific fact.

Prayer: Dear Lord, the world scorns Your Word and the history it teaches us. I ask that You would help me be a better witness to the Bible's history, for part of that history is Your work of salvation for me. Amen.

REF: Bower, B. 1990. Cyprus dig pushes back colonization date. *Science News*, v. 123, Dec. 8. p. 359

The Insect with Two Brains

2 Corinthians 5:17

"Therefore if any man be in Christ, he is a new creature: old things are passed away; behold, all things are become new."

We have all looked in wonder at the amazing transformation of a caterpillar into a moth. However, did you realize that a moth needs an entirely different kind of brain than a caterpillar? We can truly say that here we have an insect with two brains!

In the caterpillar stage, the moth lumbers around in a body that is primarily designed for eating. Its brain must be able to command powerful jaws and a comparatively huge digestive system as well as the complexities of caterpillar locomotion. The world of the caterpillar is very different from the world of the moth.

During metamorphosis the caterpillar completely dissolves and the moth is rebuilt from the liquid. The brain, for example, requires new nerve cells and new connections between brain cells, while some brain cells change function, migrate to new locations, or even die. The result is a brain that is 10 times the size of the caterpillar's and ready to control flight. The moth's primary purpose in life is to reproduce. So its new brain is able to generate the odors used by moths to identify and communicate with potential mates. It is also equipped to detect the scents of a potential mate.

Only our almighty Creator could completely rebuild a caterpillar into a moth. Only He could rebuild and increase its brain ten-fold in only a week or two. Even more amazing is how He remakes people. Through the forgiveness of sins, Jesus Christ turns us from people who wish to escape Him to new creatures who love Him and find joy in Him.

Prayer: Dear Lord, I thank You that of all the miracles You have ever worked in this world, You have worked the greatest on me by giving me faith in the forgiveness of sins You won for me on the cross. Amen.

REF: Miller, Julie Ann. 1983. A brain for all seasons. *Science News*, v. 123, Apr. 23. p. 268.

Humans Copy the Creator

Job 39:19

"Hast thou given the horse strength? hast thou clothed his neck with thunder?"

If, as we are continually told, evolution is a fact, why is it that human technology improves when we copy designs found in nature? If nature is really a giant trial-and-error experiment, we should not only find evidence of less sophisticated designs in the distant past, but humans today should be able to improve on designs found in nature.

Science has long known that an object moving through water or air moves with the least resistance when the ratio of its length to diameter is four to one. This ratio is found in the fastest swimmers in the sea—like the tuna, dolphin and swordfish. A Boeing 707, with a ratio of nine to one, experiences much more resistance than the larger, fatter 747, which has a ratio of six to one—much closer to the tuna and the dolphin.

Studies of the vision of the horseshoe crab, supposedly one of the earlier forms of life, have taught engineers how to produce a clearer, sharper television picture. The United States Air Force copied a speed sensing system from beetles to improve the ground speed indicators of their jets. The optics of a frog's eye showed them how to improve their radar. The fly's multifaceted eye taught engineers how to design a lighting system that generates more light with less energy.

Humans are learning from and copying design solutions worked out by their Creator. The brilliant designs in our world are not the result of trial and error. This reality makes the claim that evolution is a fact of science sound empty!

Prayer: Dear Lord, we humans make inflated claims about our knowledge. But in the end all that we are, have, and know is from You. I thank You for Your undeserved generosity. Let me never forget Your goodness. Amen.

REF: Man's mimicry. *Lutheran Science Institute Journal*, Jan.-Feb. 1991. p. 4.

Are Bees Intelligent Planners?

1 Corinthians 2:7-8

"But we speak the wisdom of God in a mystery, even the hidden wisdom, which God ordained before the world unto our glory: Which none of the princes of this world knew: for had they known it, they would not have crucified the Lord of glory."

Before we had detailed knowledge of the daily lives of many animals, it was easy for scientists to dismiss signs of animal intelligence as instinct. After all, to the evolutionist, intelligence can only evolve in the highest creatures—like humans.

In the last couple of decades, animal intelligence has become a busy field of study. Researchers are learning that animals can be just as intelligent as humans in their own area of specialization.

In one experiment, researchers moved a supply of sugar water 25% further away from a beehive each day. The bees quickly figured out what was going on and started anticipating where the sugar water would be the next day. In another experiment, researchers placed sugar water on a boat that was anchored in the middle of a small lake. When scouts returned to the hive to report their find, other bees refused to go with them, knowing that bees don't find food in the middle of a lake. Scientists have concluded that bees think events through to see if they make sense.

Creationists know that intelligence did not evolve. It is a gift of the Creator who generously shared what was needed with each creature He created. We also know that intelligence is not what separates us from the animals. Only humans were made for an ongoing, intimate relationship with the Creator. That also means that each of us is responsible before our Maker. God has wisdom, not just intelligence, and He wants to share the wisdom of His forgiving love through Jesus Christ with us!

Prayer: Dear Father, people are impressed with intelligence. Help me to be better able to communicate the wisdom of Your forgiving grace to us through Jesus Christ, which is far better than intelligence. In Jesus' Name. Amen.

REF: Do bees plan ahead intelligently? *Science News*, v. 123, Apr. 23, 1983. p. 271.

The Animal with Two Life Cycles

Psalm 148:7

"Praise the LORD from the earth, ye dragons, and all deeps:"

The jellyfish almost seems like a creature from a bad science fiction movie. Around 500 species of jellyfish, ranging in size from small marbles to several feet across, inhabit the world's oceans. Jellyfish have several mouths and will try to eat anything, including each other. Thankfully the stings of most jellyfish have no effect on human beings.

The jellyfish lives two very different lives, each with different lifestyles and methods of reproduction. The familiar jellyfish form is a swimming lifestyle in which the creatures reproduce sexually. However, in another phase of their life, they are called *polyps*. In the polyp stage they look like small anemones, cling to rocks and reproduce by cloning.

Jellyfish cannot move fast enough to chase prey. However, careful study of jellyfish swimming patterns shows that jellyfish don't swim at random. The pattern in which they swim produces the greatest mathematical likelihood of encountering prey. Biologically, jellyfish have only two internal organs, a stomach and sex organs. Although evolutionists tell us jellyfish were among the earliest creatures, some of them have sophisticated brain functions that—outside of jellyfish—are found only in mammals and human beings!

The jellyfish is a tribute to God's creativity and imagination. Even though the jellyfish appears to be one of the simplest forms of life on earth, it is also a mathematically precise hunter. Even this living bag of lukewarm Jell-O praises God through its existence.

Prayer: Dear heavenly Father, just as nothing in the creation fails to praise You, let me never fail to praise You before Your people, in Your house, and before the world. In Jesus' Name. Amen.

REF: Brownlee, Shannon. 1987. Jellyfish aren't out to get us. *Discover*, Aug. p. 49.

Mapping Intelligence

Psalm 119:105

"Thy word is a lamp unto my feet, and a light unto my path."

Picture yourself sitting in a friend's living room. Can you picture in your mind where the door is located? Scientists call this a thought map. Thought maps are mental pictures of the arrangement of objects relative to one another.

Thought maps help us remember where to find a water glass, the gas or brake pedal, and home. Thought maps require intelligence. They help make possible additional intelligent activity, such as finding our way around and conducting an intelligent search. This is why scientists have been surprised to find that many creatures, not just humans, use thought maps.

One experimenter carried a chimpanzee on his back while he hid 18 pieces of food in a field. Then the chimp and five others were released in the field to find the food. The chimp who had seen the hiding places found most of the food following a very different route than that originally taken by the researcher. Bees, likewise, use mental maps. Even if a bee is forced to detour on its way back to the hive, it always reports the food source as though it had used a direct route. Bees are also able to find a food source that has been moved after its discovery. Ants have also been shown to use mental maps.

Mental maps help humans and animals think deductively about changes in the environment. Mental mapping ability could not have evolved, because even the lowliest creature requires this ability to survive. This ability must have been given to them in the first place by their Creator.

Prayer: Our dear Father in heaven, I thank You that You have made me and the whole creation. Natural laws could not have made us, especially with such wisdom. In Jesus' Name. Amen.

REF: Miller, Julie Ann. 1984. Mind maps. *Science News*, v. 125, Jan. 28. p. 62.

Bees—Master Engineers

Psalm 119:103

"How sweet are thy words unto my taste! yea, sweeter than honey to my mouth!"

Why did the Creator bring together so many precise specifications and high intelligence in the honeybee? To glorify Himself and because bees are crucial to the reproductive cycles of so many plants as well as the food chain.

The amazing structure of the honeycomb has amazed scientists for thousands of years. As early as the third century AD, the astronomer Pappus of Alexandria offered an explanation for the six-sided or hexagonal shape of the honeycomb. He pointed out that while the triangle, square and hexagon could all be candidates for the shape of the comb, one of these shapes holds more than the others. The most efficient shape for storing the most honey while wasting the least space is the hexagon.

There are many variations of the hexagon that are available to the bee. However, the honeybee is always very precise in constructing honeycomb hexagons, with an obtuse angle of 109°23′ and an acute angle of 70°23′. These angles produce the most usable space for the least amount of building material. Humans are unable to use such precision in their designs without sophisticated equipment.

Each honeybee is a tiny but powerful computer that could only have been designed and built by an intelligent Creator. Bees and what they do are of central importance to much of the rest of creation. No wonder God has given them so much; the complex structure of their society, their intelligence, and industriousness!

Prayer: Father, when I look at Your personal care in the design of the honeybee, it is clear that You have a very personal love for all You have made. Remind me of this especially when I am in need of comfort. In Jesus' Name. Amen.

REF: Philip, Johnson C. 1991. Bees—Master Engineers. *Bible-Science Newsletter*, v. 29:3, Mar. p. 1.

Our Father

Matthew 5:48

"Be ye therefore perfect, even as your Father which is in heaven is perfect."

What did God have in mind when He created the human male? The world around us offers many images of what it means to be a "real man." The world offers us the bread-winner, the tough-guy and the workaholic's unending push for success. It's becoming increasingly universal that it is the woman who holds the family together. We can be sure that none of these images capture what God had in mind.

If we want to learn some practical lessons about what God intended a man to be, we need to begin with God's original creation of the first man. The first job given to Adam, the first man, was an intellectual task—identifying and naming the animals.

Jesus Christ taught us to call God our Father. However, we shouldn't make the mistake of comparing Him to our imperfect human fathers. Human fathers need to learn from God. What kind of father is He? He's tough. However, toughness does not mean hurling thunderbolts. True toughness is best shown when love persists despite rejection and disobedience. God never wearied of patiently seeking Saul of Tarsus, His sworn enemy.

However, while He is loving, God is not a wimp. He has established absolutes that we can depend on. He always keeps His Word. We can be thankful for that because when we look at ourselves we know that we have not been the men or women He intended us to be. His Word of grace in the forgiveness of sins through Jesus Christ assures all who cling to Christ that the perfect and almighty God is truly our Father!

Prayer: Our Father in heaven, I ask that Your name would be kept holy among us through the conduct of our lives and our persistent love for each other. In Jesus' Name. Amen.

Spidery Liquid Crystals

Psalm 150:1-2

"Praise ye the LORD. Praise God in his sanctuary: praise him in the firmament of his power. Praise him for his mighty acts: praise him according to his excellent greatness."

Scientists and manufacturers are studying a new technique that could lead to more durable fibers. This technique manufactures liquid crystals to produce stronger, longer lasting fibers. The process uses sulfuric acid, temperatures well above the boiling point of water and a high-pressure atmosphere.

Meanwhile, another team of researchers finally discovered how orb weaving spiders and silkworms manufacture their silk. Scientists had theorized that silk from spiders and silkworms was stronger by weight than steel because proteins in the silk cross-linked with each other. But after watching silk production and drying in careful lighting under magnification, astonished scientists actually saw what happens.

Silk is produced in the silk gland as a thin, watery liquid. Once the protein-rich liquid is released from the gland, the liquid begins to concentrate because of evaporation. As the liquid concentrates, it changes into a liquid crystal state. Scientists have been trying to create fibers in the lab from liquid crystals. They would love to know how the spider does it using only water at normal temperatures and pressures.

This is another example of how God has already solved an engineering problem that we were trying to solve. It is especially interesting that He allowed scientists to discover *His* solution at the same time that others were trying to solve the problem. This allows us to reemphasize that our world is the work of a Creator and not mindless, natural accidents.

Prayer: Dear Lord, You are Lord not only of the altar and the pew, but also of the spider, the silk manufacturing process and all that exists. Let me always remember that You are also my Lord through the forgiveness of sins. Amen.

REF: Cowen, R. 1991. Liquid crystal bridges silk-spinning gap. *Science News*, Feb. 23. p. 119.

Was Fruit Like This in Eden?

Genesis 1:29

"And God said, Behold, I have given you every herb bearing seed, which is upon the face of all the earth, and every tree, in the which is the fruit of a tree yielding seed; to you it shall be for meat."

It is difficult to imagine what things would be like had we not fallen into sin. We have never known the perfect world that God created for us.

The Bible does tell us that God gave man every fruit to eat except for the fruit of the tree of the knowledge of good and evil. We know that fruit today is highly perishable. Fruit that is picked incorrectly or stored improperly or too long quickly goes bad. As a result, we are all familiar with those green or pink flavorless things called "tomatoes." Tomatoes in the market are picked green and then exposed to ethylene gas to make them turn red. They are picked green because tomatoes quickly over-ripen. Somehow it seems to us that this state of affairs would be different in a perfect world.

Scientists at Cornell University have developed a natural way to slow the ripening process of tomatoes. They learned that there is a natural process that allows ripe tomatoes to keep for longer periods of time. They learned this from a Brazilian tomato that keeps for weeks even after being picked ripe and flavorful. This tomato has twice as much of a chemical that prevents over-ripening as other tomatoes. This natural chemical, added to other tomatoes, doubles their shelf life. This allows them to be picked ripe and flavorful. Perhaps before sin all tomatoes had this ability.

Sin touched every part of our lives. We thank God that through the forgiveness of sins in Jesus Christ, sin's effect of separating us from God is reversed!

Prayer: Dear Lord, I thank You that You have allowed us to, in part, reverse some of the outward results of sin. Let me never lose sight of the fact that only in You is sin truly reversed through forgiveness. Amen.

REF: No more pink tomatoes? *Reader's Digest*, Mar. 1990. p. 135.

The Brainy Jellyfish

Psalm 51:6

"Behold, thou desirest truth in the inward parts: and in the hidden part thou shalt make me to know wisdom."

Jellyfish and their cousins are often described as the simplest animals without nervous systems. Because evolutionary scientists have thought jellyfish were such simple creatures, and because many of them show up early in the fossil record, jellyfish were said to be among the earliest animals to evolve.

Marine biologists have been surprised to learn that the modern populations of jellyfish and their cousins are actually much more complex than they ever suspected. Creationists point out that because there are so many of these creatures, there was also probably a large population in the past. The more common a creature is, the more likely it is to be found in the deeper rocks of the fossil record.

More recently, scientists have learned that the supposedly simple nervous system of the jellyfish is actually not so simple. They have confirmed that at least one type of jellyfish—and likely more—has a nervous system with the same advanced features found only in mammals and humans. Most animals generate electrical impulses using sodium. However, like humans, this jellyfish can generate different kinds of impulses, depending on whether sodium or calcium is used. Thus, the jellyfish can send twice as much information through its nervous system as most other animals!

We're learning that when humans refer to a creature as simple, it's probably because of human ignorance of the creature, *not* because it's a simple form of life. This should remind those of us who believe in creation that all things witness in support of God's wisdom, not human stories of evolution.

Prayer: Dear Father, with Your wisdom You not only provide for Your creatures' needs, but You also put human wisdom to shame. Teach me more of Your wisdom through Your Word. In Jesus' Name. Amen.

REF: Brownlee, Shannon. 1987. Jellyfish aren't out to get us. *Discover*, Aug. p. 53.

Humans: The Artists

Genesis 4:21

"His brother's name was Jubal: he was the father of all such as handle the harp and organ."

Which is more likely to last buried in the ground for a few thousand years, a wooden spoon or a stone hammer? Since the hammer will last long after the wood has decayed, the further back we go in human history, the more likely we are to find only the most durable things humans left behind.

This general principle explains why those who believe in evolution talk about "stone age" people. They often portray so-called "stone-age" people as primitive brutes who carried clubs around, grunted a lot and were definitely in need of a bath. Those of us who know that humans have always been human know that this picture is inaccurate. It is unfortunate that, for example, the literature of our earliest ancestors, or the toys they undoubtedly made for their children, have not been preserved to our day.

Do you think it sounds strange to talk about the literature of "stone age" people? Yet modern science does recognize their art. So-called prehistoric people have left us hundreds of beautiful paintings in caves around the world. Over 220 paintings are found in Lascaux, France alone. Research reveals that the painters who created this beautiful and sensitive art did so using artificial lighting, palettes and scaffolding. They carefully milled and mixed minerals to create a variety of colors that remain bright and rich thousands of years later.

As we look at the paintings left by so-called "stone age" painters, we can clearly see that the claims that humans have progressed up from the brute animal stage are false.

Prayer: Dear Father, people would sooner see themselves as animals than as the special creatures You created for a relationship with Yourself. Help me to grow in my relationship with You through Jesus Christ, my Lord. Amen.

REF: Digging up the real story. *Science News*, Oct. 1984. p. 87.

Flying Sheep?

Romans 8:28

"And we know that all things work together for good to them that love God, to them who are the called according to his purpose."

One unusual argument that the universe is the work of the Creator rather than aimless chance goes like this: only with the highest quality research and work, often over many years, can humans begin to approach the excellence of things in nature.

One would think, for example, that it would be a simple job for modern science to create a fireproof fabric better than anything found in nature. After all, no living things in nature— besides humans—crash and burn. However, 80% of those who die in airplane crash fires are killed by the blinding, poisonous smoke, much of which comes from their seat cushions. The foam in the cushion begins to produce toxic smoke when the treated upholstery fabric gives out at 1,300° and allows the flames through.

It took two years of work before an upholstery manufacturer in Fort Lauderdale, Florida, came up with something better. The new fabric, which has since been used on some airline seats, has stood up to a 5,000° blowtorch flame for several minutes without catching fire! Amazingly, the fabric itself is a very dense weave of *wool* backed with flame resistant yarns. Its inventor says, "It's one of God's ideas … There are things about wool that are beyond what man can do."

Sheep have no need to produce wool with properties that could help protect them in a plane crash—as would have to be the case if evolution were correct. However, since sheep are part of a larger design in the Creator's provision for humans, He has given their wool qualities needed by humans as well as by sheep.

Prayer: Our dear Good Shepherd, You know and fill our needs even before we are aware of them. Help me to remember this especially when I become anxious about my material or spiritual needs. Amen.

REF: God has a good idea for airplane seats. *Discover*, July 1987. p. 7.

Coral Grows Faster than Thought

2 Peter 3:4

"And saying, Where is the promise of his coming? for since the fathers fell asleep, all things continue as they were from the beginning of the creation."

Many people know that, despite its appearance, coral is actually a tiny animal. Even a small coral growth may be made up of billions of tiny animals, each living inside its limestone skeleton.

What many people don't know is that coral cannot live without the help of algae—a plant. Thousands of one-celled algae live in the tiny stomach of each coral polyp. The polyp requires the waste products of the algae to complete its nutrition requirements and to get the calcium with which it builds reefs. In return, the polyp's stomach provides an ideal protected living place for the algae. Incidentally, the wonderful colors in living coral come from the algae in each polyp's stomach.

Evolutionists have said that it must take millions of years for the tiny polyps to build large coral reefs, like the 2,000-kilometer-long Great Barrier Reef in Australia. The Great Barrier Reef, incidentally, is considered the largest structure ever built by living organisms. However, it has now been shown that coral polyps can double their own weight in just over 10 days! As a result, the claim of millions of years for the age of the Great Barrier Reef has been reduced by a factor of 40!

Many of the long age figures claimed by evolutionists have had to be drastically revised downward as we have learned more about the creation. The resulting figures fit the Bible's account of history much more closely.

Prayer: Dear Father, I thank You that as our knowledge of Your creation grows, Your Word is being vindicated from the charges that it is in error. I pray that this would serve as a witness to bring more people to Jesus Christ. Amen.

REF: Langreth, R.N. 1990. Bleached reefs. *Science News*, v. 138, Dec. 8. p. 364.
Chen, I. 1990. Great Barrier Reef: A youngster to the core. *Science News*, v. 138, Dec. 8. p. 364.

Illogical Wisdom

Isaiah 55:8

"For my thoughts are not your thoughts, neither are your ways my ways, saith the LORD."

In 1981, it was only a theory—one that seemed to go against all logic. The theory said that if you have a signal, say a sound that was too weak to be heard, adding a little noise would make it easier to hear. If you added too much noise, you would drown out the weak signal. Add too little noise, and it wouldn't help at all. But add just the right amount of noise, and a sound signal too weak to be heard could be heard.

The theory was shown to be a scientific fact in 1983. This illogical effect was given the name "stochastic resonance." Since the 1980s, scientists have discovered numerous examples of stochastic resonance and have researched various uses for it. Radio static makes it hard to listen to a station, especially if the signal is weak. However, researchers are building unique radios that add just the right amount of static to make impossible-to-receive stations understandable.

Now for the surprise! Did you ever wonder how you are able to select a single voice in a noisy, crowded room and hear what it was saying? Scientists now believe that our brains have built-in stochastic resonance processors much more sophisticated than the units they hope to build someday. As one scientist said, "Nature may have understood stochastic resonance long before we did."

What seems illogical—to add noise to a sound to make it clear—is actually an astonishing piece of wisdom. The wisdom in the way this advanced technical knowledge is used is yet one more witness, not to impersonal "nature," but to our Creator.

Prayer: Dear Father in heaven, how little we still understand about the wonderful things You designed and created. I pray that our current age would stop talking about an impersonal "nature" and recognize You. In Jesus' Name. Amen.

REF: Peterson, I. 1991. The signal value of noise. *Science News*, v. 139. Feb. 23. p. 127.

Olber's Paradox

Psalm 148:3

"Praise ye him, sun and moon: praise him, all ye stars of light."

Why is the sky dark at night? There's more to the answer to this question than the fact that the stars do not appear as bright as the sun from earth.

If stars were evenly distributed throughout an infinite universe—as predicted by the evolutionary big bang theory—a star would be shining no matter where you looked in the night sky. The night sky, while not as bright as the daytime sky, would not be dark anywhere. The fact that this is not what we see at night has been named "Olber's Paradox," after the German amateur astronomer who first asked the question.

Ever since, astronomers have been trying to fit Olber's Paradox into whatever theory about the universe was current. Modern followers of the big bang theory have written that the main reason every inch of the night sky is not glowing with stars is that the expanding universe prevents space from filling with light. New calculations and research now show that if the universe were expanding, it would have very little noticeable effect on the stars in the night sky. Even more interesting, concluded astrophysicist Paul Wesson, is the fact that the main reason the night sky is not bright is that the universe is simply *not old enough* to have filled with light.

The daily cycle, which includes night's darkness, is important to almost all the life that the Creator placed on earth. The truth is that we live in a universe that has been designed for life.

Prayer: Lord, as Scripture says, even the night sky sings forth Your glory. Help my life to glorify You in ways that others can see. Amen.

REF: Cosmological paradox in the dark of night. *Science News*, v. 139. Feb. 23. p. 125.

Biblical Critics Disproved Again

1 Samuel 11:1

"Then Nahash the Ammonite came up, and encamped against Jabesh-Gilead: and all the men of Jabesh said unto Nahash, Make a covenant with us, and we will serve thee."

Over the last couple of centuries, many Bible scholars have developed a skeptical approach to the Bible. Their treatment of the Bible as nothing more than a human document stems from their belief in evolution. According to them, Christianity has evolved from earlier religions just as humans evolved.

As a result, critics reject the idea that the Bible is God's inerrant, inspired, trustworthy Word. Because they hold this low view of the Bible, they believe that they have found many errors in the Bible over the years. At one time they said that the Hittites mentioned in the Old Testament never existed. They said that ancient biblical writers invented these people so that Israel could make up pretend victories over them. Eventually the Hittites were discovered, one of the first evidences being hard-to-ignore six-foot-tall jars! Over the years, these critics have been consistently proven wrong by new discoveries.

Likewise, critics have said that the Israelites could not have entered Canaan when the Old Testament says they did. They reason that because the Old Testament mentions the Ammonites, which critics said didn't live in Canaan until later, the Bible's dating is in error. However, in the summer of 1989, biblical archaeologists proved that the Ammonites were in Canaan well before the Bible's date for the exodus. Again, Bible critics have been proven wrong.

As one biblical archaeologist wrote, "As usual, the absence of evidence is not necessarily the evidence of absence." When the Bible says something is true, there is no question of its truth.

Prayer: Dear Father, I thank You that Your Word is always true, because people are easily fooled and fall into error. Build me up in both faith and the knowledge of Your Word. In Jesus' Name. Amen.

REF: Late bronze city found in Jordan. *Archaeology and Biblical Research,* v. 4, n. 1., Winter 1991. p. 27.

Fake Spider

1 Peter 5:6-7

"Humble yourselves therefore under the mighty hand of God, that he may exalt you in due time: Casting all your care upon him, for he careth for you."

One of the spider's virtues is that it loves to eat flies. However, not all spiders catch flies in their webs.

The jumping spider stalks its prey until, like a tiny tiger, it is close enough to pounce. Rather than defining its territory with a web, the jumping spider takes its territory with it, in a manner of speaking. Wherever the jumping spider is becomes its territory. When one jumping spider encounters another, they wave their legs at each other, warning each other to back off. Herein lies the tale of how one species of fruit fly is able to outsmart the spider. Unlike most fruit flies, this particular species has stripes on its wings. When it waves its wings over its head, it looks something like a jumping spider declaring its territory. When stalked by a jumping spider, the fly is able to convince the spider that it's another jumping spider protecting its territory.

Researchers have discovered why this trick works even though the fly doesn't look much like a spider. The spider's response to the pattern is pre-wired into its brain and is triggered by certain cells in its eyes. These cells are only activated by a display of the pattern. The spider automatically reacts to the pattern. All the fly has to do is fake the pattern and it's safe!

The fruit fly's strategy requires a sophisticated knowledge of the jumping spider's nervous system as well as its behavior. It's safe to say that no fruit fly was able to create this strategy. It could have been authored only by the God who created them both.

Prayer: Dear Lord, You care for all of Your creatures and do so in ways that create wonder in us. I pray that Your care for me will always be such that it creates wonder and praise from me as well as others. Amen.

REF: On display: the stars of the stripes. *Discover,* May 1987. p. 9.

The Deceptive Eye

Psalm 139:14

"I will praise thee; for I am fearfully and wonderfully made: marvellous are thy works; and that my soul knoweth right well."

Do your eyes ever play tricks on you? Our eyes regularly play tricks on us with the help of our brains.

Repeated research has shown that the images we see are not exactly what our eyes see. Our eyes and brains work together to construct the images we call sight. It's not that our eyes are out to hoodwink us. Most of the time our eyes are working to help us. Working together, our brain and eyes add logical details to the images we see. Each of us has a blind spot in our vision because there are no vision cells where the optic nerve is. Instead of showing us a spot of nothing at the blind point in our field of vision, our brain invents an image from surrounding details. Your brain pulls the same trick on you when you proofread your completed manuscript. While others can still find typos, you may not. Your brain knows what you are saying and so it automatically shows you a repaired version, making many of your typing errors invisible.

Your pupils can reveal your thoughts. Studies show that when seeing unpleasant scenes, the pupils retract. Pleasant or interesting scenes cause the pupils to open wider.

Charles Darwin wrote, "To suppose that the eye, with all of its inimitable contrivances for adjusting the focus to different distances, for admitting different amounts of light, and for correction of spherical and chromatic aberration, could have been formed by natural selection, seems, I freely confess, absurd in the highest possible degree." We don't often agree with Charles Darwin, but this time we do!

Prayer: Dear Father, I thank You that the excellence of Your work in making us is so noteworthy that even those who want to deny You cannot easily do it. Teach me how to use this to lead others to Christ. In His Name. Amen.

REF: McCutcheon, M. 1989. *The Compass in Your Nose...* Los Angeles: Jeremy P. Tarcher.

A Rose Is Still a Rose

Isaiah 40:8

"The grass withereth, the flower fadeth: but the word of our God shall stand for ever."

Roses have fascinated people for thousands of years. While modern rose strains are the result of two centuries of breeding, the Greek naturalist Theophrastus wrote about the hundred-petaled rose 2,400 years ago. The Autumn Damask Rose, which blossoms in both spring and fall, was nurtured and cultivated by the ancient Romans. The first settlers in China began to cultivate roses over 4,000 years ago.

Rose breeding is one of the oldest human experiments in the science of genetics. Roses are bred for variations in size, color, frequency of bloom, and scent. More than 25 distinct scents have been identified in roses. By the mid-19th century, breeders had developed some 4,000 varieties of roses.

Perhaps roses with names like Chrysler Imperial, Charlotte Armstrong or Mirandy sound familiar to you. The man who developed those roses and has won over a dozen "All America Rose of the Year" awards is Walter Lammerts. He became one of the world's leading authorities on roses while he worked as the senior breeder at Armstrong Roses. He also developed four varieties of peaches.

Dr. Lammerts is a creationist. He not only rejects evolution, but he blames belief in evolution for slowing the development of better strains of plants. He points out that breeders who think change takes place slowly and through evolution don't expect to make much progress, and they waste time trying to fit what they learn into evolution. He adds that, despite thousands of years of breeding, roses are still roses.

Prayer: Dear Father, I thank You for the improvements to life on earth that have been made through the efforts of Bible-believing scientists like Dr. Lammerts. In Jesus' Name. Amen.

REF: Mohs, M. 1987. Where has all the fragrance gone? *Discover*, June. p. 90.

Sleep and Bacteria

John 11:11-12

"These things said he: and after that he saith unto them, Our friend Lazarus sleepeth; but I go, that I may awake him out of sleep. Then said his disciples, Lord, if he sleep, he shall do well."

When you are ill, you need sleep. And when you are sick, you are usually sleepy. Medical researchers are now learning that it's not accidental that illness and sleepiness go together. In fact, the relationship is one of God's ingenious designs.

Early in this century, scientists isolated what they called the *S factor*—the S being for sleep. Modern biochemists have identified the S factor as a biochemical that is released from the cell walls of bacteria when our bodies destroy them. This chemical finds its way to the brain where it induces sleep. It also stimulates the immune system. Normally, the body breaks down some of the harmless bacteria that live in our digestive systems every day. By the end of the day, enough sleep factor has collected in our blood to make bedtime welcome.

However, when we are fighting a bacterial infection, our body is destroying many additional bacteria and releasing a lot more sleep factor. That sleep factor is also stimulating our immune system to higher levels of activity. Researchers recommend that we give in to our desire to sleep when we are ill. Our immune system and non-dreaming sleep are closely related. In other words, grandma was right when she told us that we would avoid getting sick if we got enough sleep.

It's no accident that bacterial infections trigger the body to react, not only to rid itself of bacteria but also to strengthen itself for the fight. This ingenious design is yet another indication that our Creator does care about what happens to us. That love is why He sent His Son to win our salvation on the cross of Calvary.

Prayer: Dear Father, I thank You for good health and the abilities You have given my body so that it can stay healthy. I pray that You would also help me stay spiritually healthy. In Jesus' Name. Amen.

REF: Feeling sleepy? Shhh! Bacteria at work. *Discover*, April 1987. p. 12.

Doctor Nose

Ecclesiastes 10:1

"Dead flies cause the ointment of the apothecary to send forth a stinking savour: so doth a little folly him that is in reputation for wisdom and honour."

We don't often think about our sense of smell. While our noses are on the job all the time, we note an odor only when it is unusual. Most of the time, the scents around us contribute, unnoticed, to our overall impression of our surroundings.

It's hard to remain tense at the beach. Simply the mention of the water and waves, the sand and sun, brings back the last trip to the beach and starts relaxing a person. However therapeutic these elements of the beach are, scientists have found that the one beach fixture that is probably most responsible for relaxing us is the smell.

The smelly chemical that utility companies put in natural gas so that we can smell leaks is methyl mercaptan. This chemical is also produced by rotting meat. The human nose is extremely sensitive to this smell. We can detect as little as one 400-billionth of a gram of it in a quart of air.

Doctors have long known that their noses are helpful in diagnosing disease. A person is brought into the emergency room in a coma. It's very likely that the doctor will smell the patient's breath as he tries to discover the problem. A sweet smell could indicate diabetes. The smell of ammonia could indicate a kidney malfunction. Arsenic poisoning smells like garlic. Diphtheria can be identified by a sickeningly sweet odor, and people with the plague smell like apples. A patient who smells like freshly baked bread may have typhoid.

While we seldom think about our sense of smell, it's a wonderful gift from our Creator for which we owe Him thanks.

Prayer: Dear Lord, I thank You for the sense of smell; it adds joy to my life and helps protect me. Help me to see and thank You for those things that I have not noticed before. Amen.

REF: McCutcheon, M. 1989. *The Compass in Your Nose* . . . . Los Angeles: Jeremy P. Tarcher. p. 99.

God's Mind Is Bigger than Ours

Job 9:9-10

"Which maketh Arcturus, Orion, and Pleiades, and the chambers of the south. Which doeth great things past finding out; yea, wonders without number."

The entire universe and all its forces were planned and created out of the mind of God. Since He is God, He was not forced to make anything in a certain way.

The study of the universe and its forces is called cosmology. The night sky is a breathtaking vision to anyone who views it far from city lights. As cosmologists look into space, they see mysteries that cannot be explained. These mysteries can literally cause supercomputers to have nervous breakdowns from pondering imponderables. One would think that the cosmologist would be more likely than most to recognize the genius of God.

Researchers recently gave a supercomputer the task of calculating the results of gravitational collapse of a huge collection of matter in space. The result was what scientists call an "infinity." This means that the answer to the mathematical problem is so complex that no scientist can calculate it. Even the most powerful computer in the world is not up to the task! The team that arrived at the incomprehensible answer admits that modern science really knows very little about the nature of the cosmos.

Cosmologists know that the universe is not an accident. They expect the universe to have laws and make sense. As they learn more about the universe, they are gaining an even greater appreciation for the Mind from which the universe has come. The study of cosmology has led even some of the most hardened atheists to admit that sometimes the idea of a Creator God does not seem so far-fetched.

Prayer: Dear Father in heaven, You are so far above us in Your thoughts and power that it seems hard to believe that You desire a relationship with me. Yet I believe Your promises to me through Jesus Christ. Amen.

REF: Peterson, I. 1991. Throwing light on cosmic censorship. *Science News,* v. 139, Mar. 16, 1991. p. 148.

A Hairy Subject

Matthew 10:30-31

"But the very hairs of your head are all numbered. Fear ye not therefore, ye are of more value than many sparrows."

They are the only cells in the human body that repeatedly die and self-regenerate throughout life. They are probably the only cells in the body that inspired a Broadway play named after them. And science considers them so interesting that several branches of science are studying them. I am talking about hair cells.

As scientists are learning more about hair, they are coming to greater appreciation of, as one researcher put it, hair's bewildering biology. A Harvard molecular biologist admitted that one of the reasons we know so little about hair is that scientists have avoided studying this complex structure.

The average scalp has about 100,000 hairs growing at any one time. Each hair follicle produces hair for about two to five years, followed by a rest period of four to six months. A miniature assembly line of cells divides within the follicle, pushing older hair cells upward through the skin. Pigmentation is added as each new cell is produced. When the follicle is at rest, it shrivels and many of its cells die. However, seed cells, promising new life to the follicle, hide within it waiting for the signal to spring back to life. As humans grow older, the resting time of the follicles increases and growing time decreases, especially in men. So if you find yourself balding, you're not necessarily losing your hair. It's just that your hair is spending more time resting than it used to.

God tells us that He cares so much about each one of us that He even knows how many hairs we have on our head!

Prayer: Dear Father in heaven, I do not understand how You can keep track of so many details about so many people. Nor do I understand how You can love us so. But I trust in Your promises to me in Jesus Christ. Amen.

REF: Weiss, R. 1991. Fuzzy science. *Science News,* v. 139, Mar. 16. p. 169.

Doctor Leech

Romans 5:9

"Much more then, being now justified by his blood, we shall be saved from wrath through him."

"Dog bites man" is not an unusual headline. However, "Man bites dog" would usually get people's attention. If that's true, the headline "Leeches teach physicians" should capture everyone's attention.

Physicians used leeches in bloodletting treatments well into the 19th century. They were used to treat everything from headaches to heart disease. As the 20th century matured, the medical use of leeches became not only a memory but a symbol of how backward medicine used to be. Now, as the 20th century closes, leeches are again a part of modern medicine.

To obtain their food, leeches inject a powerful blood thinner called hirudin. This hirudin is what interests modern medical researchers in leeches. Doctors have for many years been using a blood thinner called heparin to dissolve blood clots, such as those associated with heart attacks. But researchers are finding that hirudin is ten times as effective as heparin in dissolving clots, and in addition it works directly to prevent clotting. The studies continue. However, medical researchers are convinced that the blood thinner leeches taught them how to make offers a powerful medicine. Should the possibilities presented by this modern medicine be realized, many lives that are now lost due to blood clotting may soon be routinely saved.

The Creator of the heavens and earth is so wise and generous that He can teach even modern science some new things through a seemingly simple creature like the leech.

Prayer: Dear Lord, I stand in awe at the work of Your hands all around me. Most awe-inspiring is Your work of salvation for me through the forgiveness of sins. This is the highest of all knowledge! Amen.

REF: Fackelmann, K.A.. 1991. Bloodsuckers reconsidered. *Science News*, v. 139, Mar. 16. p. 173.

Surprising Facts About the Eye

Psalm 139:14

"I will praise thee; for I am fearfully and wonderfully made: marvellous are thy works, and that my soul knoweth right well."

Our eyes do more work and tell more about us than many people realize. These amazing organs are so closely tied into the brain that they have been referred to as part of the brain.

Your eyes were among your first organs to form. Within four weeks of fertilization, an unborn infant's eyes begin to form. Interestingly enough, many newborns can move their eyes independently of each other, which can be a shock to unsuspecting parents. Usually, the eyes are working together within a few months. However, those eye muscles will get more of a workout during our lifetimes than any other voluntary muscle. During any 24-hour period, our eyes will move about 100,000 times. That's equivalent to walking 50 miles every day!

Studies have shown that people with dark eyes generally have faster reaction times than people with light-colored eyes. Brown-eyed people generally are better at tennis, while blue-eyed people generally do better at golf. Our eyes secrete different kinds of tears. Normal tears contain bacteria-fighting chemicals produced by cells in the lining of the eyelid. Emotional tears contain hormones released by the body in response to stress, including a natural painkiller. One more eye fact. The vitamin A in carrots is necessary for the proper working of the rods in your eye. The rods themselves are responsible for night vision. On a clear, dark night a healthy eye can see the light from a single candle 25 miles away!

As the Bible exclaims in praise of God, "I will praise You, for I am fearfully and wonderfully made."

Prayer: Dear Lord, I thank You that You spared nothing in making me. And when sin had taken me, I thank You that You spared nothing, including Your own life, for my salvation. Let my life thank and praise You. Amen.

REF: McCutcheon, M. 1989. *The Compass in Your Nose* . . . . Los Angeles: Jeremy P. Tarcher. p. 101.

Garlic Medicine

Numbers 11:5

"We remember the fish, which we did eat in Egypt freely; the cucumbers, and the melons, and the leeks, and the onions, and the garlick:"

Garlic. Even say the word and some people move away to give you some distance. Others gather around, because to them, garlic helps make the meal. Garlic has begun to attract the attention of medical researchers as well.

At the Tagore Medical College in India, researcher Arun Bordia kept track of 432 heart attack survivors for three years. During that time, half of the survivors drank the juice from six to ten cloves of fresh garlic each day while others took a garlic scented drink with no garlic in it. Over the three-year period, the garlic group had 32% fewer heart attacks and 45% fewer deaths than the other group. Research in the United States with chickens, who metabolize cholesterol in much the same way as humans, suggests a reason for these results. Chickens that were fed aged garlic extract showed a 50% reduction in the kind of cholesterol that clogs arteries.

Medical researchers are also looking at some interesting evidence that garlic may help prevent some forms of cancer, especially stomach cancer.

When the Creator made the heavens and earth, He knew that humans would disobey Him, destroying the creation with sin. He also knew that in doing so we would bring disease and death into the world. We can thank Him that He stocked the earth with substances that could be used to treat some of the earthly effects of sin. However, we can thank Him even more joyfully that He provided the full cure for sin for each of us in the life, death and resurrection of His Son, Jesus Christ!

Prayer: Dear Father, I ask that You would not allow the wonderful gifts You have given us to make this life better distract me from my eternal needs and Your spiritual supply in my Savior, Jesus Christ. Amen.

REF: Garlic medicine: cures in cloves? *Science News*, v. 138, Sept. 8, 1990. p. 157.

Skeletons Puzzle Evolutionists

Exodus 20:11a

"For in six days the LORD made heaven and earth, the sea, and all that in them is, and rested the seventh day..."

Contrary to what the theory of evolution says, the history of life found in the fossil record does not show simpler kinds of living things gradually becoming more complex. Nor does the fossil record show a few kinds of life gradually becoming more kinds of life. The fossil record doesn't even show one example of either of these claims that are universally accepted by evolutionists.

Life appears suddenly in the fossil record. And the living things found in the earliest rocks are completely developed and well equipped for life. Trilobites had fully formed eyes that were apparently more sophisticated than many creatures today. They appear without ancestors.

According to evolution, one of the greatest steps forward by living things was the development of skeletons. The first evolutionary step toward skeletons started with the development of calcium shells by marine creatures. Yet, the fossil record shows that creatures with calcium shells simply appear fully formed, as do creatures with internal skeletons. One question that troubles evolutionists is why creatures suddenly decided they needed skeletons. Incidentally, the astonishing variety of creatures in the oldest rocks shows a greater variety than is alive today.

Of course, the sudden appearance of fully formed, well-equipped life agrees nicely with the Bible's claim that all life was created fully formed by God. When the evidence from science agrees so well with the Bible, those who insist on evolution do so only by faith—a faith that is against the Bible.

Prayer: Dear Lord, in their pride, people rage against Your truth. Use me as Your instrument to make the testimony of Your truth more clearly heard so that fewer are lost to a false faith. Amen.

REF: When life got hard. *Science News*, v. 138, Aug. 25, 1990. p. 120.

Plant and Bacteria Communicate

Ephesians 2:8-9

"For by grace are ye saved through faith; and that not of yourselves: it is the gift of God: Not of works, lest any man should boast."

Some people talk to their plants. But what do you say to a bacterium? Evidently the soybean plant knows just what to say to make bacteria its close friend and helper.

Symbiosis means a relationship between two different creatures in which they both help each other. Plants need nitrogen for healthy growth. However, while there is more than enough nitrogen in the air, plants normally can't make use of that nitrogen. Legumes, like soybeans, link up with bacteria that are able to take nitrogen from the air and turn it into a form of nitrogen that plants, including the soybean, can use. In return, the plant creates a nice home for the bacteria in nodes in its roots and provides the bacteria with food.

The bacteria need a great deal of oxygen and energy to fix nitrogen into the soil. The oxygen is supplied by heme. If that name reminds you of the hemoglobin that carries oxygen in blood, you're on the right track. That's why the heme in the nodules turns them reddish. Who makes the heme, the soybean or the bacteria? When alone, neither creature has any heme. Researchers have learned that the plant performs the first part of the chemistry needed to make heme and the bacterium finishes the job. The plant and the bacterium actually communicate chemically with each other so that production goes smoothly!

Communication between such different creatures, allowing them to perform sophisticated chemistry to improve both of their lives, shows how inadequate evolution is to explain life. This arrangement surely glorifies the Creator!

Prayer: Dear Father, without Your communication to me in Your Word, my life would be without the cleansing benefits of my Lord and Savior's blood. I thank You that my salvation is without my effort, which could never be enough. Amen.

REF: Pennisi, E. 1991. Intimate chemistry of a symbiotic odd couple. *Science News*, v. 139, Mar. 9. p. 151.

The Flying Insect Eradicator

Ephesians 3:20-21

"Now unto him that is able to do exceeding abundantly above all that we ask or think, according to the power that worketh in us, Unto him be glory in the church by Christ Jesus throughout all ages, world without end. Amen."

For many people, bats bring to mind words like "creepy," "fangs," "rabies," and even "ugly." However, over 450 years ago, Martin Luther marveled at the delicate beauty of the bat. He pronounced the creature an excellent witness to the skill and wisdom of the Creator.

God must also like bats. Almost one-fourth of all mammals He created are bats! The flying fox bat has a wingspan of nearly six feet. Bumblebee bats, native to Thailand, are the smallest mammals known, weighing less than a penny. There are brilliantly patterned tropical bats called butterfly bats. And hundreds of species of tropical plants depend on bats as pollinators.

Popular legend considers bats a serious part of the rabies problem. The truth is that raccoons, skunks, foxes, and even dogs and cats are more likely to have rabies. In fact, one study shows that in Pennsylvania you are more likely to find a rabid cow than a rabid bat. In reality, bats are delicate, helpful and intelligent creatures. A typical North American bat will eat about 600 mosquitoes an hour. Large colonies will consume between 250,000 and 500,000 pounds of insects per night! Bats can be trained, demonstrating their intelligence. And finally, an important warning. If you find sleeping bats, don't disturb them. Even one such disturbance during sleep can be fatal, since stored nutrition may be consumed by the disturbance before the next feeding cycle.

So next time you see a bat, don't recoil in fear. Rather, say a prayer of thanksgiving to the Creator who made both of you.

Prayer: I thank You, Dear Lord, for the wonder that Your handiwork inspires. Fill me with a thankful heart that thinks of things in the same way that You think of them rather than as the world thinks of them. Amen.

REF: Givens, K.T. 1990. Going batty. *Modern Maturity*, Oct.-Nov. p. 60.

The Delaware Creation

Acts 17:26

"And hath made of one blood all nations of men for to dwell on all the face of the earth, and hath determined the times before appointed, and the bounds of their habitation;"

Many cultures around the world have legends about creation and a great Flood that have been handed down from generations in the distant past. Most of these creation and Flood stories support the biblical story of creation and the Flood.

The Wallam Olum is the story of creation and the history of the Delaware Indians. It predates the coming of white people or Christianity to North America. The Delaware creation legend reveals an all-powerful creator, who began by creating an earth without dry land, and the heavens including the sun, moon and stars. After dividing the waters from the land, he made all living things, including the first mother. This world was full of joy and had no sorrow until the snake brought wickedness. It concludes by saying that all this happened beyond the great ocean.

The Delaware story of the Flood begins with the world filled with evil because of the activity of the evil snake. While the Delaware legend blames the Flood on the snake, a few were saved through the Flood at an island called "Nana," a name very similar to Noah.

While the Wallam Olum is not Genesis, the similarities with Genesis are too many and too exact to be accidental. Similar legends from all over the globe have the same or similar details that they, too, share with Genesis. This can only be explained by the fact that the Bible is accurate in describing how all cultures have shared this common history together.

Prayer: Dear Father in heaven, fill me and all believers with a greater concern and spirit of sacrifice to do what is needed to bring the good news of salvation to those who are without it. For Jesus' sake. Amen.

REF: Branscome, Gary R. 1991. The Delaware Indians and the creator. *Bible-Science Newsletter*, May. p. 1.

Is Your Doctor an Old Crab?

Matthew 6:7-8

"But when ye pray, use not vain repetitions, as the heathen do: for they think that they shall be heard for their much speaking. Be not ye therefore like unto them: for your Father knoweth what things ye have need of, before ye ask him."

Plants make cellulose. Cellulose is a type of natural plastic made of many sugar molecules linked together. Scientists have known for years how to use the natural plastic qualities of cellulose to make things like cellophane and rayon.

Scientists have also known that animals generate a similar natural plastic called chitin. Like cellulose, chitin is a carbohydrate made of long chains of sugar molecules. Chitin is found in the shells of all insects and crustaceans, like crabs. Each year the crabmeat industry had thrown away thousands of tons of chitin because scientists couldn't figure out how to extract chitin from the shells in a usable form.

Finally in 1975, scientists discovered how to remove the chitin from shells and change its form so it could be made into whatever forms they needed. Suture thread made from chitin dissolves in the body. This means that stitches don't have to be removed, and it can be more easily knotted than regular suture thread. Suture thread made from chitin also causes no allergic reaction and it seems to promote healing. It's hard to believe that a crab shell, which is usually thrown away, could do so much! Food processing companies are even using a chemical made from chitin to recover protein from wastewater.

Almost everything in the creation seems to have numerous uses for human needs. This is testimony that the Creator has made this universe for us and that He cares about our needs. In the Bible we learn that He has provided for our greatest need through His Son, Jesus Christ.

Prayer: Dear Father in heaven, I thank You that You first loved me even when I had no good thought for You and that You have provided for the forgiveness of my sins through Your Son, Jesus Christ. In His Name. Amen.

REF: Dworetzky, T. 1987. How an old crab could keep you in stitches. *Discover*, Feb. p. 16.

Medicine That's a Spice of Life

Psalm 6:2

"Have mercy upon me, O LORD; for I am weak: O LORD, heal me; for my bones are vexed."

If you had painful skin, one of the last things you would apply to it would be hot peppers. Not only do hot peppers burn the mouth, some can even blister the skin.

Yet in South America, ground hot peppers have been used for centuries to help relieve pain. Scientists have discovered the active ingredient in peppers that relieves skin pain. This substance blocks the pain signals that come from the nerves just under the surface of the skin. The "hot pepper" sauce for skin not only causes no pain, it relieves pain that doesn't respond to any other treatment. It's available in creams and ointments that, tests show, relieve pain in two out of three people.

Conventional wisdom says that medicines that are found in natural plants were discovered by accident. Someone might accidentally discover a relationship between a person's having eaten a certain plant and then getting well. They would try the same plants on others with the same disease. If they found that there were more cures, the plant would be remembered as a folk medicine.

The "hot pepper" pain reliever for skin tells us that there's a problem with this theory. People who have a painful skin condition are not likely to apply a pepper sauce. Doing that would be expected to make their skin more painful and cause it to blister. However, it's possible that our Creator originally gave humans some knowledge of the cures for our ills—cures that He wisely built into creation.

Prayer: Dear Father in heaven, I thank You that You have provided help for our earthly sicknesses. Make me keenly aware that these cures cannot replace Christ in providing for my deepest need, the forgiveness of sins. Amen.

REF: Check, William A. Hot peppers for pain. Los Angeles Times (Repr. in Reader's Digest).

A Touching Truth

Ecclesiastes 3:5

". . . A time to cast away stones, and a time to gather stones together; a time to embrace, and a time to refrain from embracing;"

Which of your five senses is most important? Which sense could be lost with the least threat to your life?

Most people consider their sense of sight the most important. While the loss of sight is a terrible tragedy, life can go on. Lose your sense of smell, and food will be flavorless, but you can get by. Loss of hearing, too, is a great loss, but the loss can be adjusted to. And few people's lives are threatened by the loss of the sense of taste. However, if you lose your sense of touch, you lose your sense of pain and, therefore, any warning that you might be injuring yourself. Unfortunately, there are people who have no sense of touch. Sometimes they receive life-threatening injuries because they cannot feel pain.

Science has shown that touching is necessary for life. Newborn animals that are licked by their mothers have a higher resistance to disease. They are also more likely to live to adulthood than newborns who are untouched by their mothers. Chihuahua pups have a notoriously high rate of death because their mothers are often unwilling to lick them. Lambs will die after birth if their mothers do not lick and nuzzle them. Even human infants who are not lovingly touched enough will suffer from depression, poor sleep, weight loss, poor immunity and, in extreme cases, even death.

Our wise Creator has built into us the need for touch. Though we take it for granted, without touch, life would be impossible. Can you name some other gifts He has given you that you have been taking for granted? Thank Him today!

Prayer: Dear Father, You are so generous and have given me so much that I could never thank You for each gift by name. Nevertheless, I thank You for all of them, and especially for the forgiveness of my sins in Jesus Christ. Amen.

REF: McCutcheon, M. 1989. *The Compass in Your Nose*… Los Angeles: Jeremy P. Tarcher. p. 112.

Seeing Eye Fish

Psalm 9:1

"I will praise thee, O LORD, with my whole heart; I will shew forth all thy marvellous works."

There's an old saying that people seldom remember how rapidly the job was done, but they do appreciate good workmanship. However, the staggering variety and number of things created by God did not limit the elegance of His fine workmanship.

The relationship between the snapping shrimp and a fish called the goby is a good example of the Creator's attention to the finest detail. The snapping shrimp has very poor eyesight, while the goby has excellent eyesight. Both share the same burrow, which is dug by the shrimp and guarded by the goby.

The shrimp uses the goby as a blind person uses a Seeing Eye dog. Whenever the shrimp is outside its burrow, it always keeps one antenna on the goby. The shrimp stays hidden inside the burrow if the goby should temporarily swim away. When danger approaches, the goby signals and disappears inside the burrow. The shrimp is right behind him.

Neither the goby nor the shrimp can survive without each other, so pairs are established for life when both goby and shrimp are very young. In order for this system to work, both goby and shrimp reproductive systems, which differ greatly, are synchronized so that shrimp and goby youngsters are ready to pair at the same time.

This astonishing relationship shows us how our Creator's standards of excellence bring His love to every corner of the creation.

Prayer: Dear Father in heaven, as Your Word says, You have done all things very well. I thank You for the comfort of knowing that nothing in the creation is too unimportant for Your loving attention. In Jesus' Name. Amen.

REF: Yanagisawa, Y. 1990. Strange seabed fellows. *Natural History*, Aug. p. 46.

Digging Is a Snap

2 Corinthians 9:5

"Therefore I thought it necessary to exhort the brethren, that they would go before unto you, and make up beforehand your bounty, whereof ye had notice before, that the same might be ready, as a matter of bounty, and not as of covetousness."

The strange partnership between the snapping shrimp and the goby is absolutely necessary if both creatures are to survive. The nearly blind shrimp relies on the goby, a small fish, in the same way a blind person relies on his Seeing Eye dog. The goby relies on the safety of the complex tunnel system the shrimp digs.

The shrimp is well designed for digging. Once a tunnel is started, the shrimp can dig a burrow large enough for both itself and the goby within a few hours. Each of its five pairs of legs is specialized. The first pair of claws is designed for carrying sediment out of the burrow. The second pair of legs is designed to probe for and clean objects as well as scratching sediment from walls. Pairs three, four and five are walking legs that also work well for scratching material from tunnel walls. On its abdomen the shrimp has several appendages that allow it to move rapidly to protection. Other appendages can create water currents that shift sand back from the digging site.

The snapping shrimp, which is less than two inches long, can dig a system of tunnels covering several square feet within a few days. The burrow will have several entrances, as much as four feet apart. All lead to a chamber at the deepest point of the system.

The Creator has given both the goby and the shrimp different gifts that they both need to live. If they did not share their gifts with each other, neither could survive. What gifts has the Creator given you to share?

Prayer: Dear heavenly Father, I thank You that You have left none of us without a special gift to be used for our own good and the good of others. Help me to better understand what You have given me and how to use it. Amen.

REF: Yanagisawa, Y. 1990. Strange seabed fellows. *Natural History*, Aug. p. 46.

The Brain and Your Health

Philippians 4:8

"Finally, brethren, whatsoever things are true, whatsoever things are honest, whatsoever things are just, whatsoever things are pure, whatsoever things are lovely, whatsoever things are of good report; if there be any virtue, and if there be any praise, think on these things."

Medical researchers are learning that your thoughts can affect your health.

The cells that make white blood cells can be directly controlled by the brain through nerves. Using the language of hormones, the brain, as well as individual nerve cells, can direct the abilities of immune cells to kill invaders. Immune cells also communicate back to the brain using hormones. It has been shown that stress and highly emotional states can make this communication less efficient, hampering the immune system.

Macrophages are large amoeba-like cells that rebuild injured tissue and devour bacteria, viruses and anything else that should not be in wounded tissue. These amazing little cells can even eat iron filings. But chemicals produced by the body during times of stress–for example, during prolonged periods of feeling helpless–actually make macrophages very sluggish.

A word of caution. While many New Agers have selectively used some of these findings to promote their pagan ideas about medicine and health, there is nothing mystical going on here. Not only does this link between our brain and our health once again glorify the Creator, it supports biblical truth. In Philippians 4:8 the Apostle Paul advises, "Finally, brethren, whatever things are true, whatever things are noble, whatever things are just, whatever things are pure, whatever things are lovely, whatever things are of good report, if there is any virtue and if there is anything praiseworthy—meditate on these things." Not only is this advice good for the soul, it is good for the body.

Prayer: Dear Lord, whenever I feel helpless, remind me of Your wonderful promises in Your Word that You are always lovingly with me. You gave up Your life for me, so I know that You will not spare anything to bring about my good. Amen.

REF: Wechsler, R. 1987. A new prescription: mind over malady. *Discover*, Feb. p. 51.

The Shrinking Universe

Job 38:31

"Canst thou bind the sweet influences of Pleiades, or loose the bands of Orion?"

Everybody knows that the universe has been expanding ever since it was created by the big bang, right? Of course, the big bang theory of the creation of the universe doesn't necessarily agree with the Bible's account of the creation of the universe.

One of the primary so-called proofs of the big bang and the expanding universe is the fact that the starlight reaching earth seems to be a little redder than it should be. Supporters of the big bang say this is because everything is moving away from us at such high speeds.

However, the opinion of the scientific community is not unanimous. Other scientists point out that starlight could be red-shifted by nothing more than gravity. Such an effect was predicted by Einstein. The Canadian astronomer J.B. Tatum says that starlight could also be shifted toward the red if the stars and galaxies were spiraling in toward the earth at high speed—a shrinking universe! V.S. Troitkii of the Radiophysical Research Institute in Russia says he has evidence that light once traveled a million times faster than it does now. He points out that the slowing speed of light would also produce the red-shift and would also mean that the universe is shrinking.

Is the universe expanding, shrinking, or remaining the same size? The truth is, compared to the universe, we are such tiny creatures that we have no way of proving things one way or the other. Thankfully, the One who created the universe, and Who is much greater than the universe, cares about each one of us. He invites us to call Him "Father" through His Son, Jesus Christ.

Prayer: Dear Father, by its very size the universe glorifies You. But I glorify You with all my being because, despite my smallness, You did not even spare the life of Your Son, Jesus Christ, when You loved me! Amen.

REF: Astronomical Society of Canada, v. 79, 1985. p. 302.
Astrophysicics and Space Science, v. 139, 1987.

The Talk Beneath Your Feet

Psalm 96:12

"Let the field be joyful, and all that is therein: then shall all the trees of the wood rejoice."

Some of the most fascinating discoveries in recent years have to do with our growing knowledge of plant communication. New research continues to reveal even more amazing facts about the exquisite intricacy of God's marvelous work.

Naturalists have long noted that plants often choose to grow at precise spacing from each other and plants of other species. Scientists at the University of California at Santa Barbara theorized that this precision might be a result of communication between the roots of these plants. So they devised an experiment to find out. They designed a network of interlocking, partially transparent growth boxes. These boxes allowed them to watch how root growth took place in single plants, two plants of the same species, and plants of different species.

They discovered that when burro weed roots neared the roots of the creosote bush, the roots of the creosote bush slowed the growth of the nearing roots. They also discovered that burro weed roots will stop growing when they encounter other burro weed roots. This allows the plants to send their roots out where they will not compete with their own kind. Most astonishing is the discovery that burro weed roots continue to grow if they encounter other roots from their own plant. The amazed scientists noted that this means that even plants may have a sense of self.

When the Bible tells us that the whole creation praises its Creator, it certainly suggests that at least every living thing has a sense of itself as well as its Creator.

Prayer: Dear Lord, I thank You for the excellence of Your work. Help me to praise You as Creator and Savior before others so that their voices, too, may join in worship. Amen.

REF: Weiss, R. 1991. Root words. *Science News*, v. 139, Mar. 23. p. 188.

Bamboozling Bugs

Psalm 111:10

"The fear of the LORD is the beginning of wisdom: a good understanding have all they that do his commandments: his praise endureth for ever."

For centuries Asian cooks have wrapped meat, sushi and candy in bamboo bark. Though the reason for this practice seems to have been forgotten, humans evidently once knew what modern science is rediscovering about bamboo.

Chemists have succeeded in extracting the natural chemicals in bamboo bark. What they found was not one or two but four natural chemicals in the bark that inhibit bacteria, especially staph germs. These natural antibacterial agents are now under study for possible use as natural food and cosmetic preservatives. If these natural chemicals can be shown to kill bacteria and yet be harmless to people, the door will be open for an entirely new class of natural food preservatives.

The question naturally arises about how Asians originally came to use bamboo in the kitchen. Is it possible that thousands of years ago, when ancestors of today's Asians headed east from the Tower of Babel, humans knew that bamboo would help keep their food from spoiling?

Much of what we are learning today was already known to a humanity that was only a few generations removed from our first and once-perfect parents. Remember that the perfection with which humanity was created was more than moral perfection. Because Adam and Eve had a perfect knowledge of God, they also had a perfect understanding of all that He had made. While we cannot be perfect except through the forgiveness of sins in Jesus Christ, it is also true that the fear of the Lord is the beginning of all wisdom.

Prayer: Dear Lord, while humans like to pride themselves on their knowledge, compared to You and even compared to our first parents, we know nothing. I ask that You would be the center of all that I know about every subject. Amen.

REF: Keeping food fresh with bamboo. *Science News*, v. 139, Mar. 23, 1991. p. 191.

Babbling Fingers

Psalm 27:8

"When thou saidst, Seek ye my face; my heart said unto thee, Thy face, LORD, will I seek."

Infants will often look ever so serious as they babble away, as if they're actually saying something of great importance. Most of us have assumed that they do this because they see others talking and they want to join in.

The clue to the fact that there is more to speech than simple imitations came when scientists discovered that infants can babble without ever making a sound. Researchers studied five infants, two of whom were deaf and had deaf parents. In the deaf households, communication took place through American Sign Language.

Scientists found that both hearing and deaf children babbled in the language used in their homes. Hearing infants produced sounds, then syllables, and eventually meaningful words. The deaf infants did the same with their hands. At first they would make the hand signs for basic letters and numbers, stringing them together without meaning. Just as the hearing infants were beginning to offer meaningful communication by their first birthday, the deaf infants' signing had reached the same point by their first birthday.

Researchers were shocked when they realized what this means. It means that the brain has some type of unified ability built into it for learning language. This conclusion goes against everything evolution says about who and what humans are. However, it fits perfectly with the Bible when it tells us that men and women were made for fellowship and communication with their Creator.

Prayer: Lord, You are the Word through Whom I was created and Who was made flesh for my salvation. I ask that You would stir up a deep hunger for Your revealed Word within the Church and in the world at large. Amen.

REF: Hands-on babbling. *Science News*, v. 139, Mar. 30, 1991. p. 205.

Feathered Language Master

Psalm 66:5

"Come and see the works of God: he is terrible in his doing toward the children of men."

It is common for world travelers to learn at least a few words of other languages as they visit different cultures and meet different people. Not only is the marsh warbler a champion world traveler, he is also a champion linguist.

The marsh warbler is a small brown bird that spends only two months out of the year in its central European breeding grounds. It spends most of the year in Zambia. The warbler travels to its winter grounds in Africa on a route that takes it through the Middle East and the Arabian Peninsula, across the Red Sea and into northeast Africa. Despite the distance warblers travel, they frequently return to the same bushes year after year.

The new generation of warblers is still quite young when they begin their 4,500-mile trip to Africa. These young warblers learn not only the calls of the various birds in Africa but also the calls of birds along the way. Warblers have been heard to imitate more than 210 species. One warbler once imitated 76 different species in 35 minutes. The following year, when the warblers return to Europe, they offer realistic mimics of other birds for only the first three or four days before they return to their own calls. This means that at the right time of year, walking along the English or German countryside, you might hear very convincing sounds from the African countryside.

The marsh warbler is a special tribute to our Creator's inventive creativity as it celebrates in song the marvelous variety He fashioned in the bird kingdom.

Prayer: Dear Lord, as the marsh warbler praises You in its special way for the wonderful variety You have created in the bird kingdom, help me to praise You for Your forgiveness in a special way before the world. Amen.

REF: Kelsey, M.G. 1991. A well-traveled warbler's repertoire. *Natural History*, Apr. p. 6.

Juicing Up Blood

Ephesians 2:13

"But now in Christ Jesus ye who sometimes were far off are made nigh by the blood of Christ."

Blood is a biochemical marvel. Blood brings nutrition to our cells and removes waste from them. It also brings our cells oxygen for life, and serves as the pipeline for our body's chemical communication systems. Medical researchers studying hemoglobin remain baffled by it because it grabs oxygen far better than it should.

Most of us think of hemoglobin as that molecule in our blood that grabs oxygen from our lungs and carries it to cells that need it. Scientists at the National Institute of Diabetes and Digestive and Kidney Diseases in Bethesda, Maryland, have learned that this popular picture of hemoglobin is far too simple.

When hemoglobin absorbs oxygen, it changes shape ever so slightly with each added oxygen molecule. This change of shape was thought to limit the amount of oxygen a hemoglobin molecule could carry. However, researchers have found that as hemoglobin changes shape, it also begins to absorb water. One hemoglobin molecule absorbs at least 60 water molecules. The water relaxes the hemoglobin so that it can carry even more oxygen. Scientists say that the water makes hemoglobin 100 to 1,000 times more efficient. Studies continue on just how hemoglobin works. It is hoped that what is learned may someday enable scientists to develop oxygen-carrying blood substances.

Blood is a true marvel of our Creator. However, no substitute will ever be found for the shed blood of Jesus Christ for us on the cross of Calvary for the forgiveness of our sins. Only in His sacrifice can we know our loving and merciful Creator.

Prayer: Father, I marvel at how You created blood to bring us life. I marvel even more and thank You for the forgiveness of sins, which I have by Your undeserved grace to me through the blood of Christ, shed on the cross. Amen.

REF: Pennisi, E. 1991. Water boosts hemoglobin's lust for oxygen. *Science News*, v. 139, Mar. 23. p. 198.

King Solomon's Pi

1 Kings 7:23

"And he made a molten sea, ten cubits from the one brim to the other: it was round all about, and his height was five cubits: and a line of thirty cubits did compass it round about."

In 1 Kings 7:23 and the verses following we read about an engineering marvel that Solomon built for the temple in Jerusalem. Different translations call it a bath or a sea.

Solomon's sea was an impressive technological effort. It held nearly 12,000 gallons of water, more than an 18-wheel tanker truck. When filled with water it weighed almost 90 tons, not counting the 12 bronze bulls on which it stood. Made from cast metal, this huge container would have had to have been cast in a continuous pouring. Bible skeptics have tried to downplay the story of this vessel. They claim that the circumference and width measurements given in 1 Kings show that the Israelites didn't even know the value of pi.

In 1 Kings 7:26 we read that its "brim was shaped like the brim of a cup, like a lily blossom." The key lies in the shape of the vessel. It was not a cylinder. It had an outward curving rim. This means that a circumference measurement taken well below the brim could be 30 cubits. Yet a measurement of the diameter at the brim itself could give a measurement of ten cubits. The two measurements would not yield a correct value for pi, nor were they intended to. The additional measurement of ten cubits from brim to brim was included to explain the fluted brim. The fluting at the brim was more than stylistic. It would have added needed strength to the vessel.

As usual, the Bible skeptics were wrong and the Word of God stands firm and true!

Prayer: Dear Lord, Your Word cannot be broken, but many have been broken trying to defeat Your Word. I thank You that in this changing world of human uncertainty I can trust Your Word forever. Amen.

REF: Zuidhof, A. 1982. King Solomon's molten sea and (π). *Biblical Archaeologist,* Summer. p. 179.

Major, T.J. 1989. Solomon's bronze basin and π. *Reasoning from Revelation*, Oct. p. 1.

A Little Does a Lot

Psalm 94:9

"He that planted the ear, shall he not hear? he that formed the eye, shall he not see?"

Every one of your cells knows how to make all of you. Believe it or not, even your hair cells know how to make a liver or an eye. This is simply another way of saying that each of your cells has within it a complete set of genetic information needed to make your whole body.

At fertilization, every one of us starts out as a single cell. As all of our cells today, that single cell is packed with all the information needed to make our bodies. However, as that single cell divides and divides again, some cells begin to form the brain while others begin to form our skin. Some become our bones and others become a liver. Medical researchers have been trying to discover why our cells specialize like this so that everything ends up in the right place.

Since the genes in any one of your cells could make any part of your body, scientists have guessed that there must be special chemicals that command the genes to use their information about your nose to make just one nose and place it where it should be. New research now supports this theory.

Perhaps this subject is so fascinating because of the genius with which our Creator made us. However, this new research in no way rules out God's personal involvement when you were formed. Only He could have packed each of our cells with the information needed to make our entire bodies.

How do these chemicals know how to put your eyes and fingers in the right places? Truly each of us is formed by the hand of God Himself, no less than were the first man and woman!

Prayer: Dear Father in heaven, I thank You for the wonderful way in which You have knit me together and given me life. I thank You even more for giving me new and eternal life through Your Son, Jesus Christ. Amen.

REF: Getting from the egg to the chicken. *Science News*, v. 138. July 7, 1990. p. 15.

The Mystery of the Creator's Name

Exodus 3:14

"And God said unto Moses, I AM THAT I AM: and he said, Thus shalt thou say unto the children of Israel, I AM hath sent me unto you."

Different tribes and nations around the world have worshiped many things as "god" throughout their histories. But in their legend and lore, nearly every people group of the world has the idea of a true God above all gods. Sometimes He is called the "High God" or "God of the sky," but He is usually recognized as the Creator. These facts support the biblical description of human history.

Even more fascinating is the fact that the word for this God is nearly identical in Africa, Asia, the South Pacific, Australia and in the European languages. Most of the names for this God around the world are based on the root word *Di* or *Ti*. This is where the English word deity comes from as does the Sanskrit *Dyu*. The Bassa of central Liberia in Africa call him *Dyu* as well. The ancient Chinese referred to the high God as *Shang Ti*, while the Havasupai of Arizona called Him *To chopa*. *Ti* or *To* is common to both North and South American Indian names for the highest God.

Dozens of similar examples could be mentioned. These similarities among the scattered people of the world show that the idea of a true God above all gods—a Creator God—is not a human invention.

According to the Bible, it was only about 5,000 years ago that all of humanity spread across the face of the earth after a worldwide Flood. The Bible says that before that time all spoke the same language. The truth of this history is witnessed by the similar and almost universal word for the Creator God.

Prayer: Father, I ask that Your Name may be hallowed among us through true worship of You as our merciful Creator Who gave His only Son, Jesus Christ, for the forgiveness of our sins. Amen.

The Magic Box

Psalm 90:2

"Before the mountains were brought forth, or ever thou hadst formed the earth and the world, even from everlasting to everlasting, thou art God."

The modern study of genetics is said to have begun in 1908 when fruit flies were first brought into the laboratory. Since that time, fruit flies have been subjected to radiation and chemicals to produce mutations. Scientists hoped that with enough mutations they would be able to prove evolution. Researchers have produced flies that couldn't fly even though they had an extra set of wings. They have produced flies with an extra eye that couldn't see as well as regular fruit flies. But they have never produced anything but fruit flies from fruit flies.

In the process, however, researchers have uncovered some astonishing facts about the lowly fruit fly. Like all other animals and humans, fruit flies start life as a single, fertilized cell. Yet, only three hours, later the egg has split into 6,000 cells. At this stage, each of these cells has already been assigned its job in life. The eyes, legs, antenna and every other part of the fly's body is already developing.

Only in the last decade did scientists learn that this amazing process is guided by a tiny structure called the *homeo box*. The homeo box in the fruit fly consists of only 180 molecules, but those molecules direct the building of the entire fly. Scientists have also found a similar homeo box in humans.

In their attempt to prove evolution, fruit fly researchers have discovered a structure that clearly reveals the Creator's wisdom and power. In the process, they have shown that the Bible is true when it says that all living things reproduce after their own kind.

Prayer: Lord, I ask that as our knowledge of the creation grows, people would find it ever more difficult to deny You. Guide them as they search so that even those who deny You today would be drawn to the cross. Amen.

REF: Brownlee, S. 1987. The lords of the flies. *Discover*, Apr. p. 26.

Squirrels Give Age of Grand Canyon

Isaiah 44:24

"Thus saith the LORD, thy redeemer, and he that formed thee from the womb, I am the LORD that maketh all things; that stretcheth forth the heavens alone; that spreadeth abroad the earth by myself;"

Who would have thought that squirrels would tell us how old the Grand Canyon is? Yet, creation scientists tell us that the tassel-eared squirrel, who lives near the rim of the canyon in Arizona, has done just that.

Tassel-eared squirrels are found on both north and south rims of the canyon. Scientists have assumed that the squirrels were there before the canyon was formed, supposedly millions of years ago. However, since the canyon has kept the two populations apart for so long, the differences between them should show what millions of years of evolution will do to squirrels.

There are minor differences between the squirrels on the north rim and the south rim. But they're not even enough to classify the squirrels as separate species. North rim squirrels have white tails and black bellies. Squirrels on the south rim have white bellies and dark tails. But many north rim squirrels have coloration like those on the south, and many on the south rim are colored like those on the north. Creation scientist Dr. John Meyer has carefully studied these squirrels. He has concluded that the squirrels on the north and south rim are actually one population that has a complete range of fur colors.

According to a law of evolution accepted by evolutionists, there should be a much larger difference between the north and south squirrels if the Grand Canyon is really millions of years old. The conclusion must be, then, that the Grand Canyon is quite young—only thousands of years old.

Prayer: Dear heavenly Father, the story of human history as told in the Bible is confirmed all around us. I pray that You would call Your people and Your Church back to complete faithfulness to Your revealed Word. Amen.

REF: Meyer, J.R. 1985. Origin of the Kaibab Squirrel. *Creation Research Society Quarterly*, v. 22.
ICR Grand Canyon Field Study Tour Guidebook. 1990.

The Bottomless Suitcase

Isaiah 29:16

"Surely your turning of things upside down shall be esteemed as the potter's clay: for shall the work say of him that made it, He made me not? or shall the thing framed say of him that framed it, He had no understanding?"

It was first discovered in 1855, but many scientists debated its existence for another 50 years. Once its existence was finally proven to everyone's satisfaction, it was another 50 years before science had any tools to begin studying it. Finally, in the late 1980s, scientists began to learn a little about the secrets of one of the tiniest, yet most amazing structures in the body.

At the back of the eye, between the deepest layer of the retina and the cells beneath it, lies a tiny moat made up of about 10 drops of a mysterious fluid. The entire moat is thinner than a sheet of cellophane.

It seems that the clear fluid in the moat serves the surrounding light sensing tissues of the eye in place of blood, bringing in nutrients and carrying away waste. It *also* transports light-sensitive chemicals needed by the light detecting cells in the eye. *In addition*, it seems to glue the retina in place. *More than that*, the moat is rich in a growth factor. This fact makes scientists believe that the gel may also be important to repairing injuries to the retina, keeping the cells of the retina young and active, as well as helping in the growth of new cells. As one researcher said, the more they study this tiny structure the more unexpected abilities they find—something like a bottomless suitcase.

Charles Darwin, who didn't even know about the moat, was right when he said that it was impossible to believe that natural selection could have produced the eye.

Prayer: Dear Father, I thank You that I have been made through Your personal attention and love and that my creation was not the responsibility of chance and accidents. I trust Your continuing love in my life through Jesus Christ. Amen.

REF: Weiss, P.L. 1990. Eye diving. *Science News*, v. 138, Sept. 15. p. 170.

A World of Dust

Genesis 2:7

"And the LORD God formed man of the dust of the ground, and breathed into his nostrils the breath of life; and man became a living soul."

It's behind your refrigerator and under your bed. It's part of the air you breathe. While dust is common, it's not commonplace at all.

Much of the dust in our everyday lives is nothing more than tiny particles of skin, clothing, carpeting—just pieces of everyday life floating in the air. Studies of dust particles reveal that most dust is soil. However, a second common dust, even a thousand miles from the sea, is salt, driven into the wind by the waves. This salt helps form the clouds that bring us rain. The next most common forms of natural dust are volcanic eruptions and forest fires. Transatlantic winds pick up the dust of the Sahara and sweep it across the ocean where enough of it lands in Miami to sometimes give the rain a light pink color.

Not only does your home have dust from all over the world, there's probably also some from outer space. Ten thousand tons of disintegrating meteorites and comets rain down on earth every year. You can even identify some of this dust by running a magnet through the dust in a rain gutter. Most of the magnetic particles that stick to the magnet are probably from space. In addition, some dust is alive, made up of pollen, fungus spores, and even dormant bacteria hoping to land on a good place to grow.

Dust reminds us that the world is smaller than we think and that it won't last forever. Are you ready for the next world? All your preparations have been made by Jesus Christ. Find out in the Bible what He has done for you today.

Prayer: Lord, as dust reminds us that this life and earth is temporary, assure me, through Your Word—which is eternal—that I shall live with You forever. Help me to look forward to this even more than I do now. Amen.

REF: Moser, P.W. 1986. *Discover*, Nov. p. 107.

The Beginning of Faith

Hebrews 11:3

"Through faith we understand that the worlds were framed by the word of God, so that things which are seen were not made of things which do appear."

Christianity is not the only religion in the world that starts with origins. While Genesis begins the Bible with the words "In the beginning…", *Humanist Manifesto I* also begins its affirmation of faith with the humanist version of origins.

Humanist Manifesto I was adopted in 1933. The document's introductory paragraphs explain that the signers feel that Christianity is outmoded. They write that the time has come, in their opinion, to establish a new religion. In closing their introduction to their statement of belief for their new religion, they write, "To establish such a religion is a major necessity of the present. . . . We therefore affirm the following:"

Fifteen affirmations follow. The first five—fully one-third—deal with origins. Genesis starts with a Creator creating all things. The humanist affirmations begin by saying, "Religious humanists regard the universe as self-existing and not created." The four following affirmations go on to affirm faith in evolutionary processes that finally produced humans.

We learn from *Humanist Manifesto I* that Secular Humanism is, by its own claims, not just another religion. It actively seeks to be in competition with Christianity. But the most important lesson we can draw from this is that religion must begin with origins to explain what human beings are. This means that the correct view of origins is essential to Christianity. If we are to remain Christian, God's account of our beginnings in Genesis must be taken seriously!

Prayer: Dear Lord, build my faith and understanding through Your Word and then let my developing eyes of faith see Your handiwork in the world even more clearly than I do now—so that I may show others! Amen.

Knowing the Religions of Our Times

Deuteronomy 12:8

"Ye shall not do after all the things that we do here this day, every man whatsoever is right in his own eyes."

What is a religion? You might be tempted to answer that it is any kind of belief about a god or some gods. However, that's not complete enough to cover all religions. There are many old religions—not to mention the new ones—that don't believe in any gods at all in the traditional sense!

Some religions even claim to be nonreligious because they reject any idea of God. Did you know that the United States Supreme Court has recognized the beliefs of Secular Humanism as a religious expression covered under the constitution's protection of religion? All of which is to say there are many religions around today, some of which we forget are religions. And some pretend they are not religions.

If we do not see ourselves as creatures, we cannot recognize the Creator. If human beings don't recognize the Creator, they are destined to try to take His place. If His values are not central in our lives, our own values will rule. If His definition of who and what we are is not part of our lives, we will make up our own definitions.

This is exactly what many are doing today. There is no shortage of religious leaders to encourage them, no matter how outrageous their beliefs. As faithful children of our Creator, we need to understand the religions of our day. We need to know what people believe so that we can show them how Jesus Christ answers their questions better than some guru or popular writer. The Great Commission gives each Christian the need to know the religions of those with whom we come into contact.

Prayer: Father in heaven, I thank You that You have made us religious creatures. Teach me how to use this in telling those around me about Your revealed truth and love for them in Jesus Christ. Amen.

Informed Christians

Matthew 10:16

"Behold, I send you forth as sheep in the midst of wolves: be ye therefore wise as serpents, and harmless as doves."

Is ignorance really bliss? There are many painful facts in our world today. We have all had times when we just wanted to turn off the evening news and not hear any more about the ways that people found to do each other in.

Many Christians find comfort in the words of the Good Shepherd who calls His sheep. We often think of ourselves as sheep in a world full of wolves who would, were it not for our Heavenly Father, tear us to pieces. However, have we forgotten that our Good Shepherd has also told us to be wise as serpents? In part, that means we need to be informed Christians.

That's one reason the Bible is such a large book, yet with *one* central message. There is so much to learn about that message. There are so many ways in which God uses His Word to touch people's lives. There are so many lives that He wants you and me to touch.

Many who will not hear His word from our lips will see our lives—and lives that follow Christ deliver a message. The worse the world looks, the more hopeless the world's news makes people. The greater the hopelessness, the clearer the message of the Christian life is.

Yes, God still does care about all the people He made. Millions of people today feel locked into purposeless lives controlled by what they think is the mindless chance that created them. You and I need to know what *they* believe. We need to be informed Christians so that we can make the witness of His truth meaningful to them!

Prayer: Dear Lord, when I am tempted to feel overcome by the evil in the world, gently but firmly remind me that because You are at my side, I have nothing to fear. Amen.

Questioning Christian Youth

Proverbs 22:6

"Train up a child in the way he should go: and when he is old, he will not depart from it."

Could it be true that we have unwittingly been telling our young people that Christianity is irrelevant?

As I present the evidence for creation and explain to young people the problems with evolution, a similar pattern has emerged with each presentation. While the information is new to them, it rings true. Soon the questions start and they begin to realize that they have been misled by those teaching evolution. Before long, though I have not been encouraging resentment, many of them begin to feel angry that they have been misled by their teachers.

When I have finished, many of them surround me to ask more questions and make comments. The words of one young man sum up why they are so vitally interested. He told me, "I just wish that my older brother had heard this message a couple of years ago. I talked with him a lot and I just know that if he knew the Bible had intelligent answers, he would not have left the church and hurt my parents so. In fact, until I heard you, I was ready to follow him."

Other young people tell similar stories. They grew up learning about the creation and Adam and Eve in Sunday school. As they were exposed to evolution, they desperately wanted the Bible to offer an intelligent alternative to evolution. No one was there to offer it to them. Our silence seemed to confirm what evolution suggests—Christianity is outdated. Could it be that through our young people the Lord is telling us that we cannot be neutral or ignorant about the question of origins?

Prayer: Dear Lord, give Your people a stronger desire to address the questions being asked today with the answers provided by Your Word so that more may see that the Bible offers an intelligent alternative to the thinking of the world. In Jesus' Name. Amen.

"Eve?"

Genesis 3:20

"And Adam called his wife's name Eve; because she was the mother of all living."

Are all human beings descended from one woman? That subject has been a popular ongoing topic in news reports and magazine articles. Research supposedly shows that science has proved that all humans living today are descended from one woman. Could this be the Eve of Genesis?

You need to know, first, that this research was produced by evolutionists to support the idea of evolution. The researchers believe that a genetic material called mitochondrial DNA provides a clock that measures evolution. They studied 147 women from different parts of the world and compared their differences. Adding up those differences and how long evolution supposedly would take to produce them from one woman led them to a figure of about 200,000 years.

The scientific problems with this approach were ignored. Other possible interpretations of the data were ignored. For example, their conclusions stretch a tiny bit of assumed fact to cover 700,000% more ground than the original facts! Obviously this claim is nothing to be too sure about. None of the popular reports on this work told readers that other researchers have shown that mitochondrial DNA offers no evolutionary clock at all!

Finally, as Christians, we know that we are all descended from Eve. However, all people today are also descended from one of Noah's three daughters-in-law. If this research were accurate, this three-point heritage would also show up. This research is not what it appears to be. We should be careful not to endorse this.

Prayer: Dear heavenly Father, You made us with such love and hope. Yet, knowing we would sin, You were ready with Your plan of salvation for us through the forgiveness of sins in Jesus Christ. Thank You! Amen.

REF: Major, Trevor J. 1990. Has Eve been discovered? *Bible-Science Newsletter*, Dec. p. 11.

The Changing Age of the Earth

2 Peter 3:5,6

"For this they willingly are ignorant of, that by the word of God the heavens were of old, and the earth standing out of the water and in the water: Whereby the world that then was, being overflowed with water, perished:"

How old is the earth? If we rule out scriptural history, we are limited to determining the age of the earth from appearances. Appearances can be deceptive.

Let's say I wanted to compare the amount of dissolved aluminum in the oceans with the rate at which it is being delivered to the oceans by rivers. Using this method I could prove that the earth is only about 100 years old! You would doubt my results if I told you that. I could respond by checking how much iron is dissolved in the oceans and how long it would have taken the rivers to deliver it. This time I can prove that the earth is 140 years old.

That's one of the problems with trying to work out the age of the earth without a guide. There is no shortage of information and methods. In fact, there is so much information that you can't tell the sense from the nonsense!

Today, evolutionists say that they *know* the earth is about 4.5 billion years old. Back in 1900 the were *sure* that the earth was between 400 and 500 million years old. Fifty years before that, they were very *sure* that the earth was between 400,000 and 500,000 years old. In the last two centuries, scientists have been positive about ages for the earth that differ by over 1,000%!

Does this sound confusing? It is. One cannot help reaping confusion when one rejects what the Bible says. And that's true whether the Bible is talking about salvation or about history!

Prayer: Lord, when humans pit their wisdom in competition with Yours and then boast, they could not look more miserable. Help me to be more impressed with the truth of Your Word than the proud claims of foolish people. Amen.

The Mystery of Red Sirius

Job 9:7-8

"Which commandeth the sun, and it riseth not; and sealeth up the stars. Which alone spreadeth out the heavens, and treadeth upon the waves of the sea."

Space is full of mysteries. Those who think that the universe is millions or billions of years old have more mysteries to solve than Bible-believing Christians do.

One of these is the mystery of a star named Sirius B. This mystery is so great for evolutionists that it was one of the main topics of discussion at a scientific symposium at Louisiana State University in 1978.

Records of Egyptian astronomers dating back to 2,000 BC describe Sirius B as a red star. The Roman senator Cicero, writing in 50 BC, also said Sirius B was red. And Seneca described Sirius as being redder than Mars. And in 150 AD, Ptolemy, one of the most famous astronomers in history, listed Sirius as one of the six red stars.

There can be no question that Sirius was red. Today, Sirius B is a white dwarf star. That's a problem because according to modern evolutionary astronomy, it should take at least 100,000 years for a red giant star to collapse into a white dwarf star.

The mystery of how the red giant Sirius became a white dwarf in less than 2,000 years was, and still is, the topic of hot debate. Sirius B calls into question the most basic theories about the supposed evolution of the universe. Obviously, those evolutionary ages are not nearly so well "proven" as evolutionists would like us to believe!

Prayer: I thank You, Lord, that even as the stars glorify You, they witness to the truth about the history of the universe as revealed in Scripture. Help me, too, to be a witness for You and Your forgiving love for humankind in Christ Jesus. Amen.

REF: Ackerman, Paul D. 1986. *It's a Young World After All*. Grand Rapids, MI: Baker Book House. p. 67.

Cooked Dinosaur

Psalm 74:16

"The day is thine, the night also is thine: thou hast prepared the light and the sun."

In recent years a number of news articles have announced that our sun is shrinking. Scientists have been examining records kept since 1750 at the British Royal Observatory. They have concluded that the sun appears to be shrinking at the rate of one-tenth of one percent per century.

If the sun is shrinking at this rate, it would have been twice as large as it is now only 100,000 years ago. And 20 million years ago the surface of the sun would have touched the surface of the earth! The dinosaurs, or more accurately, their descendants, would have been cooked, to say the least. No life would have been possible on earth! This obviously poses a huge problem for evolutionists.

Evolutionists wanted to escape the hot water created by too large a sun. So they suggested that perhaps the sun alternately shrinks and expands. There is evidence of a very small pulsation, but overall the shrinkage is undeniable, and at any rate of shrinkage, this places the maximum age of the earth sustaining life at only thousands of years. Some evolutionists have now suggested that the earlier studies were wrong and the sun is not shrinking at all. The matter is still under study.

Whatever results from this debate about the sun, we need to remember that God, as Creator, is in charge. Just as the sun must obey His wishes, so the few thousand years of world history that begins in Genesis can be trusted as His handiwork that serves Him faithfully for our good as well.

Prayer: Father in heaven, I thank You for the warmth and light of the sun. However, when my eyes become too focused on this life, remind me of Your Son so that my eyes may turn toward eternity with You. In Jesus' Name. Amen.

REF: Ackerman, Paul D. 1986. *It's a Young World After All.* Grand Rapids, MI: Baker Book House. p. 62.

A Janitorial Service in Space

Genesis 1:16

"And God made two great lights; the greater light to rule the day, and the lesser light to rule the night: he made the stars also."

Let's say that you drop by a friend's house, and you find her at one end of the kitchen with a sponge mop in her hands. Three-quarters of the kitchen floor is wet. That picture provides you with a type of clock. You know that she has been mopping the kitchen floor for a while and is nearly done.

It's much the same in space. There is a great deal of dust in between the earth, the moon, all the other planets, and the sun. However, you won't need to send your friend with the mop to clean it up, because the sun and planets slowly sweep this dust from space. If you have ever seen a meteor or "shooting star" streak through the night sky, the bright trail was probably made by a particle no larger than a grain of sand. The earth was doing its part in the solar system's ongoing janitorial work. Yes, some of this space dust actually ends up in our homes!

Unlike what often seems to be true in our homes, there is a limited supply of dust in space. It is, however, still a huge amount. Like your friend's kitchen floor, we might get an idea of how long the earth, the sun, and the other planets have been around by finding out how much of that dust has already been cleaned up. According to scientists' own figures, our earth and the rest of the solar system could not have been around more than 10,000 years!

This is such a mystery to scientists who believe that things are billions of years old that some have even suggested that this dust doesn't follow the laws of physics. However, this is no mystery to those of us who know that the Bible provides for only about 6,000 years of earth history—and therefore only about 6,000 years of space dust janitorial service!

Prayer: Dear Father, help me to understand and believe all that You tell me in Your Word, for here is where I learn of the forgiveness of my sins in Your Son, my Lord and Savior Jesus Christ. In His Name. Amen.

REF: Ackerman, Paul D. 1986. *It's a Young World After All.* Grand Rapids, MI: Baker Book House. pp. 29-40.

The Famous Red Shift

John 10:35

"If he called them gods, unto whom the word of God came, and the scripture cannot be broken;"

You have no doubt noticed that the pitch of a train whistle changes from the time the train is moving toward you to the time it passes you. This lowering of pitch is an example of the Doppler effect.

Light behaves the same way. If the source of light is moving away from you with enough speed and you had special equipment, you could measure the red shift of the light. You could even use the amount of red shift you measured to figure out how fast the light source is moving away from you.

Of course, you might have a sneaky friend who likes to play tricks. If your friend put a red filter or some type of strong electronic field between you and the light, you would still see a red shift. In this case, you would be making a mistake to think that the light was moving.

Evolutionists' arguments for the great age of the universe are based on the assumption that the red shift of starlight is due to the Doppler effect. But if the light is shifted toward the red by any one of several other possible causes, the whole idea of the big bang—and billions of years—is lost. At least one respected scientist has been gathering information that clearly shows that the red shift of starlight is not due to the Doppler effect. His controversial work was described in the January 1988 issue of *Popular Science.*

Once again, the case for the evolutionary view of the universe offers no firm challenge to the history of the universe as recorded in the Word of God.

Prayer: I thank You, Lord, that Your Word is certain and sure, for the word of humans is not. Without Your Word we could not know if there was any truth. Your Word is truth; keep me growing in Your truth. Amen.

REF: DeYoung, Donald B. 1988. *Astronomy and the Bible*. Grand Rapids: Baker Book House. pp. 90-91.

Does Time Work Miracles?

Isaiah 44:8

"Fear ye not, neither be afraid: have not I told you from that time, and have declared it? ye are even my witnesses. Is there a God besides me? yea, there is no God; I know not any."

It has been said that if you take a frog and turn it into a man by adding a kiss, you have a fairy tale. But if you take a frog and turn it into a man by adding millions of years, you have science. It seems that many people think if you add millions of years, the impossible becomes possible.

Even the most committed evolutionists admit that our knowledge of how things work makes it hard to explain how life could develop all by itself from nonliving stuff.

The late George Wald, who taught biology at Harvard, said just this in an article in *Scientific American.* He was marveling at life and all of its complex systems when he penned some of the most famous words ever offered by an evolutionist. He wrote: "The time with which we have to deal here is of the order of 2 billion years. What we regard as impossible on the basis of human experience is meaningless here. Given so much time, the impossible becomes possible, the possible becomes probable, and the probable becomes *virtually certain.* One has only to wait, time itself performs the miracles."

Notice how Professor Wald admits the need for miracles to produce life. Will huge amounts of time—time we know didn't exist—create these miracles? Not at all. A fundamental law tells us that if something won't happen by itself in a short period of time, it is *virtually certain* that it will not happen over a long period of time.

It's true. If you reject Almighty God as your Creator, you will find another false god to work miracles. For many evolutionists, that false god is time.

Prayer: Dear Father in heaven, because of what sin has done to our minds, we will believe some unbelievable things, yet reject the simplest truths You teach. By your forgiving grace, create in me a new mind. In Jesus' Name. Amen.

REF: Wald, George. 1954. The origin of life. *Scientific American*, Aug. pp. 45-53.

Monkeying with Typewriters

Acts 17:29

"Forasmuch then as we are the offspring of God, we ought not to think that the Godhead is like unto gold, or silver, or stone, graven by art and man's device."

Nearly everyone just "knows" that if enough monkeys were allowed to pound away at typewriters for enough time, they would eventually produce all the great works of literature. But there are some fundamental laws of science and some very simple commonsense facts that show this claim to be nonsense.

Perhaps after a day or two some monkey might type the word "*to*" with spaces before and after it. Eventually another monkey might type the word "*be*." Given sufficient time and monkeys, we might end up with "To be or not to be." By this simple, mindless chance typing of letters we could eventually get all the great works of literature.

The truth is, time works against our fearless troupe of innumerable monkeys. If they are going to get their work done, they have to get their job done in a hurry. Neither monkeys nor typewriters last forever. Then there is the problem of paper. Someone has calculated that even if monkeys and typewriters lasted forever and there was an unlimited supply of paper, the worthless trash our monkeys would produce would fill the entire universe before they could get out one Shakespeare play—not to mention *War and Peace*!

Neither science nor the laws of probability support this nonsensical theory about monkeys. Common sense shows us that this illustration, often used by evolutionists, reveals the anti-scientific nature of evolution. The Bible isn't against science. *Evolution* is antiscientific!

Prayer: Father, as St. Paul told the Greeks, You are not a product of the human mind; we are a product of Your mind. Restore to our culture a desire to seek You as You are, and grant Your people a bold witness. In Jesus' Name. Amen.

Consider All the Dating Methods

Hebrews 11:1

"Now faith is the substance of things hoped for, the evidence of things not seen."

How old is the universe? Evolutionary scientists talk about how radiometric dating methods prove the universe and the earth are billions of years old. Let's look at *all* the methods scientists could use. There are well over 100 methods available! There are thousands of scientists who prefer other methods that offer younger ages for our world and universe.

Those few methods—out of over 100—that evolutionists prefer are not very exact. Sometimes evolutionists will actually get a negative age for something using one of their preferred methods. In other words, the same method that shows that some object is a billion years old may show that some other object doesn't even exist yet!

The fact is, more dating methods prove, just as scientifically, that the world is less than 500 years old than prove it is more than a billion years old. Of over 100 methods we know of—giving results of anywhere from 100 years to billions of years—15% of all methods give a date of between 6,000 and 10,000 years for the age of the universe. That is a very powerful statistical argument for a young creation!

We Christians should not allow evolutionists to intimidate us with scientific mumbo-jumbo. Our faith is in the Bible that God has given us. Those who promote evolution have their faith in their own reasoning abilities. The origins debate is religious. The religion revealed by God is challenging, head-to-head, another religion. That other religion, evolution, was hatched in some little human mind!

Prayer: Dear Lord, it seems that we are more easily influenced by other people than we are by You. Help me to see more clearly with the eyes of faith so that I am not deceived by human illusions. Amen.

Did God Create by Evolution?

John 1:3

"All things were made by him; and without him was not any thing made that was made."

Nearly 50% of the people in the United States, including many Christians, believe that God did indeed create all things. Unfortunately, they believe He used evolution as His instrument of creation.

Christians often adopt this idea because they are unaware that there are thousands of scientists who believe God created the entire universe supernaturally in six days. Many are unaware that good scientific reasons exist to accept God's work of creating just as it is described in the Bible. While it doesn't get much media attention, the work being done by these creation scientists is challenging evolutionism. The work of creationists has appeared in scientific journals. Scientists who believe the truth of the Bible's account of creation are involved in the professional scientific dialogue that continually goes on among science professionals.

At seminars conducted by these scientists, I have heard smiling, joyful people telling everyone they saw, "I didn't know that the Bible offered an intelligent alternative to evolution. I didn't know there were so many well-educated scientists who were creationists! What I have heard today shows me that I can be a more faithful follower of my Lord Jesus Christ and give up belief in evolution!"

There are no "facts" that demand that an educated person accept evolution as fact. The testimony of thousands of believing scientists confirms this. The instrument of God's creation was not natural law, but His Son, Who took our form upon Himself in the Person of Jesus Christ for our salvation!

Prayer: I thank You, Lord, that the truth is clear enough to see and believe. I believe, Lord. Help my unbelief. Amen.

Does Evolution Allow the Supernatural?

Psalm 14:1

"The fool hath said in his heart, There is no God. They are corrupt, they have done abominable works, there is none that doeth good."

Let's be honest. If you believe God created, but that He used evolution to do so, you have to answer a difficult question. Does evolution allow the supernatural to take place?
What is the difference between the scientific evidence needed to accept that evolution occurred without God and that needed to accept theistic evolution? Practically speaking, the physical evidence for both positions should be identical.

Evolution is built on the principle that natural laws determine what happens in the physical world. Evolution rules it out of bounds for God to get involved in the workings of a reptile egg to produce the first mammal-like creature. Evolution says there is no need for supernatural intervention to modify some ape-like creatures into the first human being. As far as evolution is concerned, God's supernatural intervention could not happen.

There is no physical evidence to support the idea that God periodically intervened in evolutionary development. Nor does the Bible talk about this kind of activity.

A person might accept that God created through evolution. But this belief would not be faithful to the physical evidences that evolutionists claim to have. Nor is it faithful to the statements we find in the Bible. Evolution does not allow the supernatural action of God. Even the Bible itself cannot be God's Word if there is no supernatural action of God. Thankfully, the Bible is God's revealed Word, and it does offer you an intelligent alternative to evolution.

Prayer: Dear Father in heaven, use me and others who believe Your Word without question. Equip us to show those who mix evolution and the Bible that there is a better path of bold faith without compromise. In Jesus' Name. Amen.

REF: Bartz, Paul A. 1990. A better path than adding God to evolution. *Bible-Science Newsletter*, Sept. p. 3.

Do Christians Need Evolution?

Acts 17:11

"These were more noble than those in Thessalonica, in that they received the word with all readiness of mind, and searched the scriptures daily, whether those things were so."

Many, if not most, of the leaders of the creationist movement today were once evolutionists. Most of them once believed that God created, but that He created through evolution. When they accepted evolution, they were considered objective by many. As soon as they rejected evolution, they were considered biased by those who still accepted evolution.

Those who were once convinced evolutionists point out that there are serious objections to evolution. For example, the fossil record paints no picture of the development of living things from simple to complex. If you are a theistic evolutionist you might not believe this, but check it out. Representatives of every major phylum of animal living today are found in the deepest layers of rocks containing multi-celled fossils. They appear suddenly without any evolutionary ancestors.

Further, the fossil record offers not one undisputed connecting or "missing link" between the various kinds of creatures. Dr. Colin Patterson of the British Museum, who is not a creationist, admits in a well-publicized letter that there is not a single transitional form to link any kind of animal with another. Even the most famous so-called transitional forms have not stood the test of scientific scrutiny.

There is no single fact or group of facts that demand that the educated Christian abandon the Bible's account of creation and history. There is no intellectual reason for Christians to adopt evolutionary beliefs!

Prayer: Dear Father, I thank You that Your revealed Word is trustworthy in everything that it talks about, including history. Help me to be more regular and systematic in my study of the Bible. In Jesus' Name. Amen.

REF: Sunderland, Luther D. 1988. *Darwin's Enigma*. El Cajon, CA: Master Books.

Certain: Six Creation Days

Exodus 20:11

"For in six days the LORD made heaven and earth, the sea, and all that in them is, and rested the seventh day: wherefore the LORD blessed the sabbath day, and hallowed it."

Many Christians are willing to accept what Genesis has to say about creation with one exception. They don't want to accept the six days as literal, natural, approximately 24-hour days. Did you know that there are other important portions of Scripture that depend on those six days being literal and normal days?

In Exodus 20:11 the Lord is giving the Ten Commandments to Moses. The Lord is explaining the principle of the Sabbath in this verse. God's explanation says that man shall work six days and rest on the seventh because God Himself made the heavens and the earth in six days and rested on the seventh. Therefore He blessed the seventh day. This truth is repeated in Deuteronomy 5:13.

Surely no Christian would argue that the Ten Commandments are *poetry* to be taken *figuratively*. The Lord is confirming here that His work of creation really occupied six normal days and that He rested on the seventh normal day. If these days were actually long ages, then we should work for six long ages before we take some rest. We can be sure that this is not what the Lord has in mind.

If the Ten Commandments should not be interpreted literally because of this connection with creation, where does that leave us? Surely the creation days, by the Lord's own confirmation and commandment to us, are literal, natural, approximately 24-hour days!

Prayer: Lord, I thank You that You have given us one day out of seven for rest and worship. I thank You for the commandments because they teach me Your will. Most of all, I thank You for Your salvation so that I can look to You and not to my own obedience for assurance of my salvation. Amen.

REF: Gentry, Robert V. 1986. *Creation's Tiny Mystery*. Knoxville, TN: Earth Science Associates. pp. 12-13.

Death Before Adam?

1 Corinthians 15:22

"For as in Adam all die, even so in Christ shall all be made alive."

If God created, but He did so by evolution, it means that there were many generations and billions of deaths before the first man appeared on the scene. Does Scripture say anything about this?

Indeed it does! In fact, the whole plan of salvation rests on the fact that there was no death before the first man. This truth is found throughout Scripture, but nowhere is it more clear than in 1 Corinthians 15:22. Here we read, "For as in Adam all die, even so in Christ all shall be made alive." This means that if death did not begin with, and because of, Adam's sin, then by the same token no one is made alive in Christ. The way it is phrased in this passage, one depends on the other. If there were generations of death in the world before Adam, then death is not tied to Adam's sin. And if death is not tied to Adam's sin, then life is not tied to Christ.

In other words, if there were death before Adam and therefore generations before Adam, Christ's work turns out to be of no effect. So you see, even evolution, with God added to it, questions the Bible's message about the saving work of Christ.

The good news is that there is no fact that demands our rejection of God's account of creation. No scientific fact has been established that forces us to reject what the Bible says about the first human beings or the origin of sin. There is no fact that has been established that rules out the Bible's account of creation, the origin of sin, or death. These truths are fact and the foundation upon which the Gospel rests!

Prayer: I thank you, Lord, for Your holy life of perfect obedience to God in my place and Your suffering the full penalty of my sin so that I might have forgiveness. As You have been raised for my justification, raise me to new life in Your grace. Amen.

REF: Bartz, Paul A. 1991. Theistic evolution offers an alien "gospel"! *Bible-Science Newsletter*, May. p. 7.

Faith: Christ or Man?

Jeremiah 23:4

"And I will set up shepherds over them which shall feed them: and they shall fear no more, nor be dismayed, neither shall they be lacking, saith the LORD."

How certain are evolutionists of evolution? Of course, no believer in evolution is going to go on national television and say that he is not sure of evolution. However, within the scientific community, those who really know the facts don't always sound so convinced of their arguments.

The great philosopher of science, Dr. Karl Popper, wrote that Darwinism is not even a scientific theory because it cannot be tested. Even the prestigious British Museum of Natural History offered this admission on one of its displays. That admission, by the way, created a scandal among American evolutionists. Dr. Colin Patterson, head of paleontology at the British Museum, was responsible for this display. He wrote in his book *Evolution* that evolution is simply a metaphysical system used to arrange scientific evidence. It is not a *conclusion* of science at all.

Even evolutionists Stephen J. Gould and Niles Eldredge recognized that the fossil record shows that creatures appeared suddenly and fully formed—without any ancestors. However, they claim that these were lucky jumps of evolution.

Our conclusion must be that even the experts, in moments of honesty—when they know they won't be quoted in the press—admit that evolution is no more than a faith. We creationists will report what the press does not. Evolution is truly a faith. We have a faith, too. But our faith has been revealed by God in His Holy Word!

Prayer: I ask, Dear Lord, that more people would realize that evolution is not a fact of science, but a faith, and a false one at that. I especially ask that You would protect Your people from being misled by this lie. Amen.

Thinking God's Thoughts After Him

Isaiah 55:8-9

"For my thoughts are not your thoughts, neither are your ways my ways, saith the LORD. For as the heavens are higher than the earth, so are my ways higher than your ways, and my thoughts than your thoughts."

Our Creator's work reflects some of His incredible inventive creativity. In Matthew 6:30, Jesus reminds us how God provides for all the needs of all His creatures. Each creature is perfectly suited for its life, even if that requires special creativity from God.

Many of the creatures that live in the deepest and darkest parts of the sea are equipped with lights. One has a very bright headlamp to light its way. Others have lights on their tails, jaws or sides. The light they produce, as the light produced by the firefly, is a cold light. The light is generated when chemicals manufactured by the creatures are mixed together.

Or consider the small bird that has an "engine" powerful enough to allow it to fly for 10,000 or more miles. The construction of the bird's heart differs from yours and mine in important ways. However, we couldn't get by with the bird's heart just as it couldn't get by with ours.

While it seems strange to think of the penguin as a bird that flies under water, that's exactly what it does. The penguin can reach speeds of 30 miles per hour underwater, as fast as the fastest marine mammal, the dolphin.

All of us would benefit from looking at the creatures around us a little more closely. If we do, we will learn a little more about the magnificent creativity of our God. As we learn about their lives and how they are provided for, we will, as one early scientist put it, learn to think God's thoughts about the natural world after Him!

Prayer: Lord, I understand that I do not think as well as or in the ways that You would have me. Create in me a clean heart and renew a right spirit within me so that I may be more conformed to Your image. Amen.

The Archer Fish

Job 12:7-8

"But ask now the beasts, and they shall teach thee; and the fowls of the air, and they shall tell thee: Or speak to the earth, and it shall teach thee: and the fishes of the sea shall declare unto thee."

The yellow-and-black striped archer fish shops among the insects that land on branches overhanging his pond in the peaceful waters of the East Indies. Sometimes he will wait under a branch until something that would make a good lunch lands on it.

When the archer fish finds a tasty-looking insect, he shoots an accurately aimed stream of water at his prey, knocking the insect into the water. Before the insect has a chance to recover, the archer fish has eaten him and is looking for dessert.

The archer fish has a deep groove in his mouth. When he places his tongue against this groove, he ends up with a blow tube that is about 1/16-inch in diameter. Compressing his gills forces pressurized water through the blow pipe. The rounded tip of the archer fish's tongue acts as a valve so the fish can shoot a few small bursts or one long stream of water at his prey. The archer fish seldom misses if his prey is within four feet.

This simple but elegant arrangement is testimony to a wise and skillful Creator. Neither mindless chance evolution nor any fish could be smart enough to design this system. Nor could any fish redesign his normal mouth tissues into this unique tool. Here we see that our Creator even cares about providing the needs of a fish. We know that He cares even more about our lives! In fact, He has said so!

Prayer: Father, You invite me to call you Father because You care for all that You have made. You proved Your love for me when You sent Your only Son to take my punishment for sin. Let me never get over my gratitude. In Jesus' Name. Amen.

REF: Coleman, William L. 1979. *Listen to the Animals*. Minneapolis: Bethany House. pp. 36-37.

The Crocodile Bird

Genesis 32:10a

"I am not worthy of the least of all the mercies, and of all the truth, which thou hast shewed unto thy servant;"

How would you like to be a dental hygienist for a crocodile? That's how the crocodile bird makes his living.

Of course you do need to know that the crocodile bird doesn't show up to do his work with any drills or needles. He and the crocodile are on good terms with each other. After eating, the crocodile climbs the river bank and relaxes with his mouth open. The little crocodile bird enters the crocodile's mouth to clean up the scraps that are left. While the crocodile bird makes most of his living as sort of a crocodile dental hygienist, he also helps keep the crocodile free of pesky insects that lodge in his skin.

The crocodile also receives one other service from the crocodile bird. Whenever the bird senses approaching danger, he gives his sharp warning call and flies off. The crocodile, now warned, can quickly roll over into the water where virtually no animal can get the best of him.

This is but one of many unlikely cooperative arrangements that we find in the plant and animal kingdoms. Every one of these relationships speaks for a Creator and against the idea that either these creatures, or their cooperation, evolved naturally. Evolutionists have written whole books on the subject. Yet they don't seem satisfied that they have explained how these relationships could develop through evolution. We agree that they have no explanation. Nor are they likely to find one as long as they deny a Creator who cares for His creation.

Prayer: Father, because of the innocent suffering and death of Your Son, Jesus Christ, I know that Your love for me is certain and sure. I thank You for this. In Jesus' Name. Amen.

Complex Ant Society

1 John 4:19

"We love him, because he first loved us."

Imagine a huge living creature covering about one square mile and made up of almost a third of a billion cells. Imagine each of those living cells being able to move about independently of the others.

No, it's not the plot of a new science fiction movie. Some scientists have suggested that ant colonies may be thought of as one large living organism. They suggest this because of the way individual members within ant colonies behave. Ant colonies have many kinds of members, each with its own specific task. The soldier ants in a colony have extra large heads and heavy-duty jaws for fighting. Other ants are concerned only about finding and returning food. Still other workers do nothing but process and store food.

Within the colony another caste of ants does nothing but tend the eggs. Others prepare the food for the larvae and deliver it to other specialized ants who do nothing but feed the next generation.

If this sounds complex, imagine the largest known ant colony. This Japanese ant colony has an estimated 306 million worker ants, and more than a million queens. The colony occupies 45,000 interconnected underground nests that range over about a square mile!

There is obviously intelligent design and control behind ant society—intelligence not found in ants or in nature itself. This intelligence comes from *outside* nature. Here is yet another witness that no concern of any living thing is too small for God's complete attention!

Prayer: Dear Father in heaven, I thank You for Your love. Help me to show love like Yours to those around me, especially those who seem least lovable. In Jesus' Name. Amen.

The Loving Crocodile

1 John 4:16

"And we have known and believed the love that God hath to us. God is love; and he that dwelleth in love dwelleth in God, and God in him."

You might not think of the crocodile as a particularly loving creature. In reality, the crocodile is among the most gentle and loving of parents. Mama crocodile lays between 40 and 80 eggs in a large nest. For the 90 days they take to hatch, she will keep close watch over them, even foregoing food so as not to leave the nest unattended. Even the father stays close to guard the eggs that are a delicacy to many animals.

At the end of this three-month watch, the babies begin to chirp from within their eggs. Before long there may be dozens of babies chirping from the nest of buried eggs. This brings the mother who carefully uncovers the eggs, one by one.

The young are only nine inches long and weigh just four ounces. Yet the gentle and loving mother picks up each one and with a flip of her head places them in a special pouch in her mouth.

When some of the young have trouble breaking out of their shell, mama picks up the eggs and delicately works the eggs inside her mouth with her tongue, freeing the imprisoned baby. Once collected, all the young are taken to a special protected nursery where they will be just as carefully tended for months. Most amazing is that, should the mother not be present to tend to the hatchling chores, the father offers the hatchlings the very same help and care!

God loves even the crocodile! And his acceptance of you and me, because of what Jesus Christ has done for us, is an even greater wonder than the loving crocodile!

Prayer: Dear Lord, Your love surpasses my understanding! Help me never to switch from relying totally on Your saving work for me to relying on my own goodness or on religion for my salvation. Amen.

Is There Life in Space?

Romans 8:22-23

"For we know that the whole creation groaneth and travaileth in pain together until now. And not only they, but ourselves also, which have the firstfruits of the Spirit, even we ourselves groan within ourselves, waiting for the adoption, to wit, the redemption of our body."

Back in 1836, well-known astronomer Sir John Herschel wrote a series of articles for the *New York Sun* reporting his discovery of life on the moon. He told of how he saw buffalo, goats, cranes, pelicans, beaches, forests, and even winged batmen. The *Sun* finally admitted that the whole story had been a prank, and most of the public was amused.

There is as much evidence today for life on the moon or other planets as there was in 1836. As we explore those other planets of the solar system, the evidence grows that there is little hope that other life will be found in space. Evolutionists, of course, expect life to be found in space. They feel that if life evolved on earth it certainly should have evolved elsewhere, too. At a popular conference on whether there is life in space, one of today's most famous astronomers, Dr. Robert Jastrow, stated that the question was "essentially a religious controversy."

That the question of life in space is a religious question is an important insight. Many evolutionists see the discovery of life in space as the final nail in the coffin of Christianity and proof of their religion of evolution. It would not be, of course.

The Bible does not offer any clear statements on the matter, although most Bible-believing scholars conclude that the Bible implies that material life exists only on earth. However, our exploration of space has taught us that our earth is a very special place, carefully designed just for life.

Prayer: Dear Father, I thank You for an earth so beautiful that even after the destruction caused by sin, the beauty around us still makes us think of You. Help me to add my words to that witness. In Jesus' Name. Amen.

REF: The great moon hoax. *Science Digest*, Nov. 1985. p. 80.
DeGarmo, S. 1984. A religious controversy. *Science Digest*, Jan. p. 8.

Real Sloth

John 9:4

"I must work the works of him that sent me, while it is day: the night cometh, when no man can work."

Some creatures are so bizarre that they suggest a lesson to us. Yet even in their strangeness we see that they have been perfectly provided for by their Creator.

One such example is the three-toed sloth. It rarely moves, spending about 18 hours a day sleeping upside down in a tree. When it does bother to move, it is in ultra-slow motion. It is not unusual for the three-toed sloth to take 30 seconds to shift one leg a few inches.

The three-toed sloth doesn't do well on the ground because he can't walk or even crawl. If a sloth is placed on the ground it v-e-r-y s-l-o-w-l-y pulls itself along on its belly to a tree. However, sloths hang effortlessly in the trees even when asleep. Their grasp of the tree is so solid that they continue to hang in the trees even after they have died.

Even though sloths are hunted for food by the South American Indians, the three-toed sloth is the most abundant tree-dwelling mammal in South America's forests. Despite the seeming helplessness of the three-toed sloth, it is doing well because it has been perfectly designed for its life. Because its life doesn't seem to amount to much, it was named after one of the so-called seven deadly sins. This creature, that has a face only a mother could love, presents an image with which neither child nor adult would want to be associated. Yet the sloth still has God's care. The nearly helpless condition of the sloth reminds us that we are completely helpless when it comes to our relationship with God. Only because of the efforts of our Savior, Jesus Christ, do we have peace with God. Let's not be slothful about telling others of the forgiveness of sins!

Prayer: Dear Father in heaven, I thank You that Your love knows no bounds. Energize me as I do Your work. Do not let me become filled with sloth or become uncaring about Your work. In Jesus' Name. Amen.

REF.: Wilder, R. When laziness pays off. 1983. *Science Digest*. p. 112.

Mysteries of the Piranha

Psalm 50:10

"For every beast of the forest is mine, and the cattle upon a thousand hills."

When the piranha is mentioned, nearly everyone thinks of a picture of a herd of cattle crossing the Amazon—and splish, splash—nothing is left but bones. Actually, the piranha is a rather interesting fish. It certainly seems to be the bearer of a few tricks the Creator wanted to play on those who want to explain the world without Him.

According to evolutionary theory, new species evolve when there is some way of making a living in nature that no one else is using. So, when you have many similar species, each one should have its own specialization. However, while there may be a dozen different species of piranha living in a portion of a river, they all eat the same things and live in the same ways. As one evolutionist put it, "This clashes with evolutionary theory."

We who believe in a Creator can understand His creating variety just for the sheer joy and exhilaration of creativity. As a result, we are not surprised to find different varieties of the same kind of fish all making the same living.

Even when piranhas do stage one of their mass attacks for which they are famous, each one takes a bite, then waits in line until the others take their turns. Many evolutionists have difficulty explaining this, too. Selflessness is not supposed to exist in an evolved world.

Could this selfless behavior be left over from when the creation was perfect? It certainly is a provision by God to make life a little better for the piranha.

Prayer: Lord, I thank You for the joy that fills me as I consider the wonderful variety of creatures You have made. Help me to better learn how to show others Your hand in the creation so that they may be led to Your forgiving grace. Amen.

REF.: The piranha's fishy feeding habits. *Science Digest*, Oct. 1983. p. 29.

Creative Procreation

Genesis 1:27

"So God created man in his own image, in the image of God created he him; male and female created he them."

While the Bible teaches us that sex is a wonderful gift of God, evolutionists believe that sexual reproduction should not have evolved. According to evolutionists, if sexual reproduction were to develop, evolution is supposed to get rid of it.

To make matters worse, immobile plants use sexual reproduction. This creates a problem in transferring the male genes to the female for fertilization. Just to confound things even more, God uses insects and even mammals to transfer the pollen. Since evolutionists believe that plants were around for many millions of years before animals, evolutionists have yet another problem.

The European family of the Arum lilies attracts beetles and flies. When one of the insects lands on a flower, he finds a slick oil-coated surface and ends up sliding down into a chamber filled with small hairs. Imprisoned, he finds a sweet sap that makes good eating. While he is gorging himself on the sap and getting sticky in the process, the male part of the flower showers him with pollen. The next day the sticky, pollen-coated insect finds that the doors to his prison are open. So he flies off to another flower to unwittingly deliver his load of pollen!

It sounds like science fiction to say that a flower figured out how to do this. And it certainly doesn't sound very scientific to say that *nobody* designed and made this complex arrangement. As we learn more about these arrangements in nature, it becomes even more clear that an all-wise Creator designed these beautiful, symphonic relationships between different members of the living kingdom.

Prayer: Dear Father, I thank You for the gift of sex. Help me to always use this gift as You intended so that my thoughts, speech and actions honor You. In Jesus' Name. Amen.

REF.: Batten, Mary. 1983. Sex & plants. *Science Digest*, Oct. p. 59.

The Strange Behavior of Orchids

Romans 1:20

"For the invisible things of him from the creation of the world are clearly seen, being understood by the things that are made, even his eternal power and Godhead; so that they are without excuse:"

There are between 10,000 and 15,000 species of orchids. While you surely are not familiar with all of them, it sounds strange to talk about orchid behavior. This huge variety of orchids offers a multitude of different flower shapes, colors and sizes. Each unique shape, color and size is perfectly suited to the bees, butterflies, bats, moths, flies and birds that are involved in that orchid's pollination.

Those orchids that use birds as pollinators have no scent. Instead they offer the brightest of colors, appealing to the birds' strongest sense. Evolutionists call this *"co-evolution."* However, giving it a name doesn't explain it. How did these orchids know that the bird's strongest sense is sight? And how were these orchids able to perform genetic recombination on themselves so that they could develop their bright colors? No, giving something a scientific-sounding name doesn't make naturalistic claims any more scientific.

Or consider the bee orchid. It offers a flower that looks like a female bee. The purpose here is to attract male bees to a position that will load them with pollen. How can it be scientific to suggest that a plant purposely restructured itself to fool bees into pollinating it?

God has purposely made it impossible to explain the creation without Him. He did this not simply so that we would see that He is real. His purpose is much greater. God seeks a personal relationship with you, and has made that possible through the saving work of His Son, Jesus Christ, for us. It is to Christ that He would ultimately draw our attention.

Prayer: Dear heavenly Father, help us to more clearly see Your power and Godhead in the creation so that we may point it out to others. Use us to bring the witness of Christ to them. In Jesus' Name. Amen.

REF.: Batten, Mary. 1983. Sex & plants. *Science Digest*, Oct. p. 59.

Three Million Children!

1 Peter 5:6-7

"Humble yourselves therefore under the mighty hand of God, that he may exalt you in due time: Casting all your care upon him; for he careth for you."

There are five known groups of orchids that throw their pollen. These flowers are designed with mechanical flower structures that work as triggers. Some types use a bees' back to trigger the pollen discharge. Another uses, yes, the bees' knees.

The *Catasetum* orchid produces a few male flowers that have a strong scent that attract a male bee. When the bee lands and trips the pollen trigger, it is showered with that flower's entire hope for the next generation. Then the flower stops producing any scent and begins to wither and die. Now it's up to the bee to find a female flower for the pollen.

It's not easy for the bee to find a female flower, because the *Catasetum* orchid produces even fewer females than males. But the female flower has an even stronger scent than the male, and it lives longer. If the bee finds a female flower, the future of this orchid is bright, because the fruit that results produces 3 million seeds! That's as if nearly half the population of New York City had the same mother and father!

Because the designs of the Creator are carried out amid such beauty and with such careful planning, evolutionists often talk about evolution as if it were a living being. However, the truth is clear. God is not merely a distant God up in His heaven. Think of it for a moment. If He is so concerned about such detail even among plants and insects, how much more concerned is He about the details of our lives!

Prayer: I thank you, dear Father, that part of Your greatness is that You are not too great to give personal care to all of Your creatures. Help me to remember this when events make You seem distant. In Jesus' Name. Amen.

REF.: Batten, Mary. 1983. Sex & plants. *Science Digest*, Oct. p. 61.

The Vine that Keeps Pet Butterflies

John 15:5

"I am the vine, ye are the branches: He that abideth in me, and I in him, the same bringeth forth much fruit: for without me ye can do nothing."

Two groups of tropical vines rely for reproduction on the world's only known species of pollen-eating butterfly. Neither vine has very showy flowers, nor do their flowers have any scent. However, they don't need to attract pollinators, because the *Heliconius* butterfly depends on the pollen produced by the vines for its food. So the butterflies don't go far from their food source.

Since the butterfly depends on the pollen, the vines produce many more male flowers than female. This keeps its army of pollinators well-fed. Being messy eaters, the butterflies end up with pollen all over them.

But what would attract the butterfly to the female flower, which can be up to 20 feet away? Two tricks are used. First, a female flower usually grows close to where a male flower has just withered. Second, the female flower, while producing no pollen, looks just like the male flower to invite the butterfly's attention.

Did you see the intelligence in this arrangement? If only chance were involved, why would the only known species of pollen-eating butterfly be found with the only known vines that rely on their services? This is a very big planet! If there is no design in this arrangement, why should these vines offer female flowers that look like the males so that they will be pollinated? It is so clear that these butterflies were custom-made for these vines, and the vines were made for the butterflies. This is a witness to a personal Creator Who is personally involved in life on earth!

Prayer: Father, just as You provide every living thing with its needs for life, You have provided for my earthly life. You have also provided for my eternal life through the forgiveness of sins in Jesus Christ. Thank You! Amen.

REF.: Batten, Mary. 1983. Sex & plants. *Science Digest*, Oct. p. 61.

Is It Orchid or Insect?

Exodus 15:11

"Who is like unto thee, O LORD, among the gods? who is like thee, glorious in holiness, fearful in praises, doing wonders?"

There are at least four known types of orchids that effectively pretend to be female insects. They are so good at this ruse that they attract the attentions of male pollinators. That there are *four* different varieties that do this makes it clear that this trickery isn't due to chance. The Creator does not want us to make the mistake of thinking that He doesn't exist.

The Ophrys group of orchids are found around the Mediterranean. Ophrys orchids produce flowers that look and smell so much like the insect that pollinates it, that the male insect will mate with the flower. In the process he picks up the pollen. The pollen is then carried to another flower, thus ensuring the next generation. The flowers even mimic the sheen of the female's wings, the furry fringe on her abdomen, and her antennae.

The *Cryptostylus* orchid, found in New Guinea and Australia, mimics the female of the wasp that pollinates it. *The Cryptostylus* does this so perfectly that it even moves as a mating female. The same is true of an orchid found in the Andes that is pollinated by a particular species of fly through the same ruse.

God has much more in mind than just letting us know that He is. He wants us to see that He is intimately involved with His creatures. Through the perfect life of Jesus Christ, and His suffering, death and resurrection, He has removed the final barriers between us and Himself. With the barrier of sin removed by forgiveness, He may be even more intimately involved with each of us on a personal basis.

Prayer: I thank You, Lord, for Your creativity and genius that is so clearly displayed in the creation. Grant me Your Holy Spirit so that I might be enabled to display more genius and creativity as I serve You. Amen.

REF.: Batten, Mary. 1983. Sex & plants. *Science Digest*, Oct.

Plants that Fool Insects

1 Thessalonians 5:8

"But let us, who are of the day, be sober, putting on the breastplate of faith and love; and for an helmet, the hope of salvation."

Many plants, especially orchids, offer different ploys to fool insects into pollinating them. Each of these ploys shows planning, design, and the ability to create the design that will work. That there are so many examples makes it impossible for any rational person to believe that mindless, impersonal forces could have made these plants.

Four different groups of orchids produce flowers that look so much like the female of the insect that pollinates them that pollinators will try to mate with them. The Brassia orchid found in southern Florida produces a flower that mimics the prey of its pollinator, the spider wasp. The flower is so designed that as the wasp stings its supposed prey—really the flower—pollen is deposited on his head. The pollen is then carried to a nearby female flower and deposited in just the right place during another attack.

Another orchid imitates an aggressive swarm of bees as it blows in the wind. The strongly territorial bees that live in the area swarm and attack the flowers. In the process, they transfer the pollen from male to female flowers. The female flowers of at least 50 species of Mexican orchids offer a fake pollen to lure pollen-eating insects that have been feeding on male flowers. Many of these insect-orchid relationships are so tightly interlocked that neither could survive without the other.

This speaks of a planned creation. It also bears witness that all forms of life came into being in a relatively short time, not over millions of years as evolutionists would have it.

Prayer: Lord, I thank You that You have so carefully and lovingly planned the creation. Help me to remember the love and wisdom of Your plans, especially when my faith is challenged. Amen.

REF.: Batten, Mary. 1983. Sex & plants. *Science Digest*, Oct.

Bamboo's Message to the World

Judges 13:18

"And the angel of the LORD said unto him, Why asketh thou thus after my name, seeing it is secret?"

Many plants reproduce simply by cloning. The most obvious example of cloning that comes to mind is the dandelion. Unless you get all of the main dandelion root out of the ground, each piece of the root can sprout another dandelion, identical to the parent.

The walking fern reproduces by cloning too. It moves across the countryside, albeit slowly, by producing new plants. A new plant is generated wherever the tips of its leaves touch the ground. Each new plant is identical to the parent.

The bamboo reproduces through underground stems. The different species of bamboo live anywhere from 20 to 120 years. During their life spans they produce only by cloning. Then they flower—once. When it's time for a particular species to flower, every member of that species, no matter where in the world it grows, flowers gloriously. Each flower can have as many as 100 pollen-producing stamens. Once the seeds are ripe and scattered, every member of that species in the world dies. At this point the entire species exists only in the scattered seeds.

This is brilliant testimony to a recent creation of all bamboos at the same time. If evolution produced them and the forces of change worked on them for millions of years, one would think that all members of the same species would no longer be in perfect synchronization. Bamboo makes a strong testimony for the recent creation reported in Genesis!

Prayer: Dear Father in heaven, I am continually amazed at the wonderful variety of life You have created. Help me to see more clearly how the creation witnesses to You and then voice that witness to those around me. In Jesus' Name. Amen.

REF.: Batten, Mary. 1983. Sex & plants. *Science Digest*, Oct. p. 61.

Can Bacteria Think?

1 Corinthians 1:25
"Because the foolishness of God is wiser than men; and the weakness of God is stronger than men."

It sounds silly to ask whether bacteria can think. However, science has known for more than 100 years that the little guys can indeed think. Experiments in 1883 conducted by Wilhelm Pfeffer showed that bacteria will swim toward good food like chicken soup and away from poisons such as mop disinfectant.

Pfeffer also learned that bacteria can make decisions. He made sure that his bacteria knew the location of chicken soup. Then he separated them from it with a mild mixture of disinfectant. He found that the little fellows would swim as fast as they could through the disinfectant to get to the soup.

This is the same type of decision-making process you and I go through every day. We often tolerate the unpleasant to arrive at the pleasant. As a result of this research, scientists today talk about bacteria actually making decisions.

These conclusions amaze most people. That's because we have been trained to think of intelligence in an evolutionary context. The "higher" or more evolved a creature is, the smarter we expect it to be. However, if we recognize, as the Bible says, that all life is the product of an intelligent Creator, we should not be surprised to find that intelligence has nothing to do with evolution. Every creature has been given as much intelligence as it needs by a Creator Who truly cares for every living creature— even bacteria!

Prayer: I thank You, Lord, that You give wisdom to the weak. Though the world exalts its own wisdom, help me to more clearly see how foolish my "wisdom" is, and how wise Your "foolishness" is. Amen.

REF.: Pietsch, Paul. 1983. The mind of a microbe. *Science Digest*, Oct. p. 69.

Bacteria's Eyes and Ears

Exodus 15:2

"The LORD is my strength and song, and he has become my salvation: he is my God, and I will prepare him an habitation; my father's God, and I will exalt him."

Modern researchers are learning that bacteria are not a simple form of life. Rather a bacterium is a tremendously complex creature that thinks. Scientists have discovered that bacteria actually have molecule-sized sensors. Some of these sensors act as eyes, while others act as ears. A dozen other proteins have been discovered that receive the information gathered by the bacteria's "eyes" and "ears." The collected information is then processed the same way your brain gathers and makes sense of information.

Bacteria have specialized senses and a brain. In fact, bacteria are better at sensing some things than you and I are. A bacterium can, for example, sense the difference between two parts and one part in 10,000.This is the same as if you could tell the difference between one jar with 9,999 pennies and another jar with 10,000 pennies!

The idea that bacteria are just "simple" forms of life comes from evolution. Growing research is showing that even one-celled creatures are not simple at all. Think of it. Every single-celled bacterium must accomplish, within that one cell, all the tasks we accomplish using the trillions of cells in our bodies. Eating, digestion, metabolism, waste removal, reproduction— and even thinking—all take place within a single cell! There is no such thing as a simple form of life.

Science certainly argues against the idea that life began by accident. What science is learning urges us to think about the Wise and thoughtful Creator Who made us!

Prayer: Dear Father, You are so wonderful that even bacteria glorify You! Excuse the weakness of my worship. Help me to praise You with all of the wonderful being You have given me. In Jesus' Name. Amen.

REF.: Pietsch, Paul. 1983. The mind of a microbe. *Science Digest*, Oct. p. 69.

The Mind of a Bacteria

Psalm 119:130

"The entrance of thy words giveth light; it giveth understanding unto the simple."

Researchers are learning that bacteria have senses similar to our hearing and sight. Bacteria even have a brain that receives information from their senses. They can make decisions. However, all of this is not enough to explain how bacteria can make the decision to swim toward food.

When you want a candy bar, you know where to find one, even if you must go to the nearest store. You might also hope that your dentist doesn't see you buying candy. All of this involves memory, which has long been considered a basic part of the mind.

Researchers, using various solutions of substances that attract or repel bacteria, have now learned that bacteria do indeed have both long-term and short-term memories. They remember where goodies are to be found. Once a bacterium develops a strategy for dealing with the unknown, it will remember what to do next time.

This finding shocked researchers. They know that when you or I decide to retreat or advance, at least hundreds and probably thousands of brain cells are involved. The one-celled bacteria can make the same decision based on past memories. Are there really "simple" forms of life that serve as the living traces of evolutionary development? Scientists haven't found any yet. According to the Bible's account of creation, we don't expect them to find any simple forms of life from which we can be traced.

Prayer: Lord, You do things so well that we often mistake Your genius for simplicity. Give me a mind more like Yours, having both love and wisdom. Amen.

REF.: Pietsch, Paul. 1983. The mind of a microbe. *Science Digest*, Oct. p. 103.

The Personality of Bacteria

Psalm 138:6

"Though the LORD be high, yet hath he respect unto the lowly: but the proud he knoweth afar off."

Scientists are learning that even the lowly bacteria have mind and memory. New research is showing that mind, memory, brain, and personality did not evolve. These can be found in the lowliest of creatures, including the bacteria.

Another sign of intelligence in a life form is the ability to learn. Using attracting and repelling substances, scientists have found that bacteria can learn about their surroundings and draw conclusions. Follow-up tests showed that they later remember what they learned. Scientists have even found that younger bacteria learn better than older ones. If some things are not learned when young, they never can be learned! It seems that it's hard to teach an old bacterium new tricks!

The biggest surprise of all came when scientists discovered that bacteria have what the researchers called individual personality. They are mystified about how bacteria, each with the same genetic structure and from the same environment, can develop unique quirks in behavior. Because we find individual personality even in the lowly bacteria, we know that did not evolve.

That God would give even a bacterium the gift of personality shows us something about His mind toward us. Nothing is so small or unimportant that it escapes His attention. Since He sent His Son, Jesus Christ, to rescue us from the decay that is in the world because of sin, there is no force too great for Him to overcome. Surely He is a God of details as well as a God of love, Who is concerned about every individual among us.

Prayer: Lord, even Your creation shows how You are concerned for every living thing. Never let me doubt Your concern about every detail of my life, and help me to communicate Your care to those around me. Amen.

REF.: Pietsch, Paul. 1983. The mind of a microbe. *Science Digest*, Oct. p. 103.

Hydraulic Spiders

Psalm 119:73
"Thy hands have made me and fashioned me: give me understanding, that I may learn thy commandments."

Hydraulic machines are part of our everyday lives. When you step on the brakes in your car, brake fluid, which cannot be compressed, is forced from a cylinder into the braking mechanism attached to your wheels. Hydraulic mechanisms work because fluid cannot expand, but it will easily flow to areas where there is less pressure.

However, humans were not the first to invent uses for this principle. The spider uses it all the time. The spider has his skeleton on the outside of his body and his muscles inside his skeleton. Since our muscles are outside our skeleton, we can use muscles both to bend our legs and stretch them out. However, the spider's arrangement allows his muscles only to bend his leg. When he wants to straighten out his leg to take another spidery step, he pumps fluid into his leg and the leg joint hinges open.

The North American dwarf mistletoe builds up hydraulic pressure equal to that found in a truck tire. It uses that stored pressure to catapult its seeds up to 50 feet at a speed close to 60 miles per hour. The squirting cucumber, found in the Mediterranean area, uses the same principle to propel its seeds up to 40 feet.

Who would believe that the brake system in an automobile came together by chance and natural forces? Neither does it make sense to believe that spiders, other animals, and plants came about without a Designer and Creator.

Prayer: Dear Lord Jesus Christ, I thank You that You have enabled us to learn how to use some of the principles You placed into the creation. Help me to understand that knowledge is never complete until I know of, and trust in, Your forgiveness of my sin. Amen.

REF.: Tributsch, Helmut. 1983. How spiders stretch. *Science Digest*, Oct. p. 74.

Dancing Bees

Deuteronomy 6:5

"And thou shalt love the LORD thy God with all thine heart, and with all thy soul, and with all thy might."

Bees depend on a good memory of their surroundings to make the thousands of trips necessary to keep their hive in operation. Each day a single bee must make trips to hundreds of flowers, some as far as a mile from the hive. Scientists have long wondered how the tiny bee can remember what may be dozens of square miles of territory.

When a bee finds a good source of nectar, she returns to the hive and does a dance for the other bees. This dance tells them where a new find is located. Researchers have studied this in detail and can now interpret these dances.

Researchers from Princeton University tricked a bee into dancing out a location that turned out to be in the middle of a nearby lake. After receiving the message, the other members of the hive simply ignored their fellow worker. She kept dancing, but no one would believe her. Only later, when the dance clearly said that the flowers were on the opposite side of the lake, did the other bees pay attention and check out the flowers.

Researchers are now wondering whether bees keep a mental map of their surroundings. How else could they tell when someone gives them a false location? Evolutionary researchers are trying to figure out how such an "evolutionarily simple" creature as the bee could be so intelligent. However, the Bible makes clear that intelligence is not related to evolution. It is the gift of an all-wise Creator.

Prayer: I thank You, dear Father, for the intellect You have given me. Help me to develop it to Your glory and service. Help me to remember that my intellectual development can never take place apart from Your Word. In Jesus' Name. Amen.

REF.: Horton, Elizabeth. 1983. Bees as mapmakers. *Science Digest*, Sept. p. 90.

Spiders the Size of a House?

Romans 12:1

"I beseech you therefore, brethren, by the mercies of God, that ye present your bodies a living sacrifice, holy, acceptable unto God, which is your reasonable service."

The largest known beetle in the world is the aptly named goliath beetle. This African beetle can become almost as large as a mouse. Dinosaurs were not the largest animals ever to live. The great blue whale holds that honor. The smallest mammal in the world is smaller than many insects. A newly discovered species of bat in Thailand weighs only 6/100 of an ounce. Such extremes might lead some to wonder whether those giant insects we see in the movies would ever develop. Could a spider grow to be as large as a house?

You will be happy to know that no spider or any other insect could grow as large as a house. We can be sure of this because of the way insects breathe. Insects breathe through small tubes along their bodies. These tubes provide oxygen directly to their internal tissues. If an insect were to grow too large, these tubes could not get enough oxygen to the tissues. This is because the larger the creature, the more internal living tissue that requires oxygen. You and I have lungs that provide a network of surfaces equal to the size of a tennis court. An insect our size would need a body that has the surface area of a tennis court, too, which is an impossibility.

Even though different creatures have different organs to supply their needs, each creature has just the right organs for its needs. This is a sign of our Creator's wisdom. This fact also tells us that living things are not related to each other by evolution.

Prayer: I thank You, dear Lord, for my eyes, ears, all my senses, and all the wonderful tissues You have built into my body for life. Help me to use my life in Your service. Amen.

REF.: Wilder, Rachel. 1983. Secrets of size. *Science Digest*, Sept. p. 112.

Animals Don't Need Technology

Psalm 147:9

"He giveth to the beast his food, and to the young ravens which cry."

When we compare ourselves with many of the other living things on this earth, our abilities are not very impressive.

A camel can carry 1,000 pounds 30 miles through the desert every day. As a human being, I become tired even thinking about walking 30 miles with empty pockets. The ostrich can run at 30 miles per hour, faster than my car might go during rush hour. The Indian swift can fly at more than 100 miles per hour. The elephant's trunk has 40,000 muscles. This is 70 times the number of muscles in your entire body. Yet his trunk can bulldoze a tree or pick up a pin. Many of us can't do either one on the first try.

The flea can jump 350 times its own length, a total of 13 inches. This is equivalent to our jumping 2,000 feet into the air. Most of us can't jump more than two feet off the ground. As it works out, 13 inches is just about right if you are looking for a nice warm dog on which to raise a family. A tiny insect called the midge beats its wings an amazing 133,000 times per second. That's a hundred times faster than you can blink your eye.

Humans can use their superior intellect to build the machines they need. With those machines, they can outdo any one of these creatures. But God simply made a gift to the animals of the specialized abilities they needed. God expects us to follow His example and design and build what we need. If we look carefully, we will see that our Creator has given us all the materials and all the ideas we need within the creation. No one can say that He does not care what we do with our lives on earth.

Prayer: Father in heaven, I am reminded again that You do care what I do with my life. Help me see Your will each day and give me Your grace to follow Your will. In the name of Jesus, Who is perfect before You for me. Amen.

REF.: Animal athletes. *Science Digest*, Sept., 1983. p. 20.

Discovery Confirms Creation Claims

Matthew 16:2

"He answered and said unto them, When it is evening, ye say, It will be fair weather: for the sky is red."

One good test of whether a theory is scientific is whether the theory can accurately predict new and surprising findings. If you have a barometer you may do this all the time. The sky may be clear, but the barometer is falling fast. No matter how nice the weather looks, you know that rain is likely.

Creation scientists believe that humans have been on earth in their present so-called "modern" form ever since the model first came out. Since that was on day six of creation, humans have been on earth as long as any creature. So creation scientists have predicted that as we find more fossils, we will find fossils of "modern man" in older and older rocks. Such a discovery would shock the evolutionists!

Evolutionists say that, according to their theory, "modern" humans have been around for only a small part of the world's history. In their inflated estimate, they say we have been here for only 35,000 years. Evolutionists received a surprise when that age recently had to be almost tripled. A new discovery pushed the date for "modern" humans back 92,000 years. As a result, evolutionists had to admit that we turn out to be older than some of the creatures from which we were supposed to have evolved!

Of course, one has to take the hot air out of those evolutionary dates. They can be adjusted down to a number that fits the biblical calendar. However, the scientific prediction made by creation scientists has proven out. And creationists are predicting that the history of "modern" humanity will continue to be pushed back further and further!

Prayer: I thank You, dear Lord, that Your Word is true and that You have preserved it to our day and age. Forgive me for not studying it as I should and help me to study as I should, beginning today! Amen.

A Desert Full of Frogs?

Psalm 51:11

"Cast me not away from thy presence; and take not thy holy spirit from me."

Even though many frogs live much of their lives out of the water, they need water to reproduce. This can be a problem, since some frogs live in trees and some even live in the desert!

For a tree frog, the trip from the treetops to a nearby pool is a long and dangerous journey. However, the Creator cares about all of His creatures. The leaves of the bromeliad that grows in the branches of many trees in the tropical forest form a private pool far above the ground. As a result, mama frog can raise her young without ever leaving the treetops!

The desert would seem to offer an even greater challenge for a frog. However, since it pleased the Creator to place frogs in the desert—maybe to challenge the evolutionists—He has also provided for their needs. It may rain only once every two years in the deserts of central Australia. When it does rain, the water-holding frogs emerge from their underground hideaways to lay their eggs in the puddles. Before the puddles disappear, not only do the eggs hatch, but the young grow and develop into frogs that are able to store water until the next rain. As the desert again takes over, the frogs dig themselves back into the ground and go into suspended animation until the next rains come. A year or two later, the new generation will emerge after the next rain to raise their young.

Neither chance nor the smartest frog in the world could design or create these abilities. Only a wise Creator could. And if the Creator has provided this much care for frogs, just think of how much more He cares about you and me in our everyday lives!

Prayer: Though the world seems very evil, dear Father, and I sometimes wonder where You are to let such evil go on, I am again assured of Your presence. Help me to communicate Your love to those around me. In Jesus' Name. Amen.

REF.: Wilder, Rachel. 1983. Flying frogs. *Science Digest*, Nov. p. 120.

A Whale of a Problem

Genesis 1:21

"And God created great whales, and every living creature that moveth, which the waters brought forth abundantly, after their kind, and every winged fowl, after his kind: and God saw that it was good."

According to evolution, life began in the sea. However, because the land was better for life, evolution developed land life from sea life. It was on land that mammals evolved, according to evolution. Here, though, evolution has painted itself into a corner. Or perhaps it is better to say that evolutionists have fallen into one of the traps set for them by the Creator. You see, the largest creatures on earth live in the sea—and *they* are mammals!

The great whales are mammals like many land animals. However, it is very clear that they never lived on land—nor do they resemble any land creature. Whales provide clear evidence that, contrary to evolution's claims, not even all mammals are related to each other. And if mammals are so different that they cannot be related to each other through evolution, neither is it likely that bugs and trees and cows and humans could be related by evolution. As Genesis so clearly tells us, living things were originally made in distinct kinds, and they continue to reproduce after their kinds.

One wants to ask the evolutionist, if the drive to live on land was so great it moved sea creatures to become land creatures, why did some mammals go to all the trouble to move back into the sea? To say the least, whales offer Bible-believing Christians a great testimony to our Creator and a good argument against evolution. For evolutionists, mammals who live in the sea are a whale of a problem!

Prayer: Your majesty and power are truly seen to all in the magnificent whales You have created, Lord. I ask that we would be successful in preserving these great creatures who witness so grandly to You. Amen.

REF: Bartz, Paul A. 1984. It takes more than this to make a whale transition. *Bible-Science Newsletter*, Feb. p. 1.

A Whale of a Tale

Romans 1:21, 22

"Because that, when they knew God, they glorified him not as God, neither were thankful; but became vain in their imaginations, and their foolish heart was darkened. Professing themselves to be wise, they became fools."

Evolutionists say that mammals evolved on land. However, there are many mammals, like whales, living in the sea who could never live on land. The story of how that happened, as told by evolutionists, shows just how much trouble evolution has explaining even simple facts.

According to evolutionists, there were (oh, about 45 million years ago) cow-like creatures who lived by the seashore. Now these cow-like creatures liked to eat fish. As they fished, the cow-like things with the biggest tails could swim the best, catch the most fish, and have more babies. Over millions of years, their legs disappeared, since they weren't using them, and their tails kept getting bigger. Their noses moved to the top of their heads, and finally the first modern whale swam the seas.

This story sounds like one Dr. Seuss made up! We have never seen such weird behavior in any cow-like creature. And not even one bone from any such creature has ever been found. Yet this is called a "scientific" theory. It certainly is hard to imagine how evolution's story about the whale can be classified as science. Science would seem to clearly say that whales were created according to their own kinds and not from some fish-eating cow!

That's where things stand today. The story of the fish-eating cow remains nothing more than a whale of a tale!

Prayer: Dear Father in heaven, I thank You that Your gifts to those who trust in Your Son for their salvation include a new mind and godly wisdom. I see the foolishness of those who lack this. Help me to grow in Your wisdom. For Jesus' sake. Amen.

REF: Bartz, Paul A. 1984. It takes more than this to make a whale transition. *Bible-Science Newsletter*, Feb. p. 1.

Whales: Engineered for Water

Psalm 119:140
"Thy word is very pure: therefore thy servant loveth it."

We are told by evolutionists that the whale evolved from a cow-like creature that originally lived on land. Let's look at that claim. First, the whale can hear what direction a sound is coming from under water. Neither cows nor humans can do this. Nor is the whale's ear damaged by the pressures deep in the sea. The whale also has other, very complex differences from land-living mammals that allow him to cope with the great pressures in the depths of the sea.

If the whale evolved from some cow-like mammal, the entire head would have to be restructured. What we could call the whale's nose is on top of his head, in back of his brain. Land-living mammals, of course, keep their noses on the front of their faces, in front of their brains. Evolutionists would have us believe that in the evolution of the whale, his nose, or at least his nostril, migrated up his face through his brain, and finally ended up where it is today. That's major restructuring!

These are just a few of the miracles we are supposed to believe that mindless evolution performed, even though it can't think or plan. And that's the operative word—*believe*. The evolutionist's faith in the evolution of the whale shows us that the origins controversy is not science versus the Bible. This debate is about a man-made and sometimes silly religion versus faith in God's revealed Truth!

Prayer: Lord, I say with the Psalmist, "Your Word is very pure." Help me to understand it and believe it, despite the arguments raised by those who value their own words more highly than Yours. Amen.

REF: Bartz, Paul A. 1984. It takes more than this to make a whale transition. *Bible-Science Newsletter*, Feb. p. 1.

Tool-Using Animals

Genesis 1:27

"So God created man in his own image, in the image of God created he him; male and female created he them."

Evolution devalues humanity. It tries to find the most unflattering way to describe human beings to try to counter the Bible's claim that they were created in God's image.

For generations evolutionists said that the humans' use of tools was all that made them different from animals. Creationists who objected, saying there was obviously more of a difference than that, were ignored as unscientific. It wasn't so long ago that evolutionists scoffed at creationists who showed them that there are animals who use tools—and, therefore, the evolutionists' definition was wrong.

Elephants often take sticks in their trunks to scratch their backs. Galapagos woodpeckers use cactus spines, held in their beaks, to probe insects out of cracks in trees. Chimpanzees even use sticks as levers to move heavy objects. They also make tools out of sticks by shaping them to the right size to poke into termite nests. The stick full of termites they pull out of the nest offers them a tasty snack. West African chimps actually use stones as hammers to crack nuts

So today, evolutionists recognize that some animals not only use tools, but also make them! Only now they claim that animals who use tools *prove* their theory. Which only goes to show that evolution is not science that can be disproved. It is a faith that will be believed even when the facts contradict it. The facts say that there is an unbridgeable difference between humans and animals.

Prayer: Lord, I thank You for making animals. I ask that You would help all who are working to teach Christian young people that they are Your creations and not glorified animals. Show me my part in filling this need. Amen.

REF.: Tool-wielding beasts. *Science Digest*, Aug. 1983. p. 107.

A Tuft Mystery for Evolution

Psalm 29:2

"Give unto the LORD the glory due unto his name; worship the LORD in the beauty of holiness."

Those of us who accept our creation as the work of the Creator find many examples of what seems to be His wonderful creativity—creativity just for the sake of creativity! These add to our appreciation of the creation and stir us to learn more of the wonders of His creative thought.

The evolutionist has robbed himself of the opportunity for that wonder. For the evolutionist, everything must have a reason related to survival. Some evolutionists even study why people think flowers are beautiful, while we enjoy flowers' beauty and fragrances.

The fact that many rabbits and hares have a tuft of longer hair in their upper chest is one mystery to evolutionists. This mane seems to be present in both the males and the females of one family of rabbits and hares all through the year. Science articles are full of speculation about how this mane might help in the struggle for survival. However, no one has been able to offer a convincing argument. Finally, someone suggested that the mane may have no survival value at all. However, evolutionists are so committed to their faith that the magazine *Science Digest* called this a "lame conclusion."

The Creator has given His creatures many things to help them survive. One of those things that He knows makes life better, but for which evolution cannot account, is sheer beauty. Let's not forget to thank Him for that, too!

Prayer: I thank You, Lord, for beauty. Help me to see Your hand more clearly in the beauty around me, so that I may praise You more boldly before others. Amen.

REF.: Sitwell, Nigel. 1983. The ruffed-rabbit mystery. *Science Digest*, Aug. p. 20.

The Most Complex Structure

Psalm 111:10

"The fear of the LORD is the beginning of wisdom: a good understanding have all they that do his commandments: his praise endureth for ever."

What is the most complex structure in the universe? I'll give you a hint: it weighs only a few pounds, and you own one.

The most complex structure in the universe is the human brain. It has so many interconnections within it that if it evolved, its evolution—even over billions of years—would be easy to measure. The fact that there is no hint that the brain is evolving is a good argument against evolution.

Intelligence is not what sets humans apart from animals. Rather, it is the fact that we were made in the image of God. The Clark's nutcracker, a small bird found in the American southwest, has an incredible memory. It can store up to 30,000 seeds for winter food in different locations. Over the winter months it will remember where the seeds are stored in all these thousands of locations so it doesn't run out of food.

It is calculated that a human brain can store up to 100 trillion bits of information. The most powerful supercomputer today can't store even 1/10th of one percent of that. That is not to say that the proper use of our thinking abilities can happen without helpful instruction. In fact, your brain's manufacturer, God, has provided just such a manual in His Holy Word. And just like a sophisticated computer, we all risk ruining our thinking abilities if we operate our brains without our manufacturer's helpful instructions.

Prayer: Dear Father, I thank You for my brain and its wonderful abilities. Help me in my use of Your Word so that your Spirit may instruct me in true wisdom and the right use of my brain in Your service. In Jesus' Name. Amen.

REF.: Finn, Robert. 1983. Amnesia: case of a fragmented past. *Science Digest*, Nov. p. 72.

The Weaverbird

Genesis 1:28

"And God blessed them, and God said unto them, Be fruitful, and multiply, and replenish the earth, and subdue it: and have dominion over the fish of the sea, and over the fowl of the air, and over every living thing that moveth upon the earth."

Intelligence—as much as is needed and no more—is generously found throughout the creation. So we cannot say that intelligence alone makes humans special.

Consider the weaverbird. The weaverbird nest consists of woven strips of fiber and grass. Using beak and feet, the male weaverbird uses both loops and knots to weave his hanging nest. Then the nest must be inspected by a prospective mate. If she doesn't like the nest's construction, she will turn down the hopeful mate. The male must then tear down his work and start over. Some males have been observed constructing and tearing down their nests two dozen times before finding a prospective mate who is satisfied with his work. Some weaverbirds actually build huge cities of nests protected by a woven roof. One roof over a weaverbird city was 15 feet across!

Human intelligence spans much more than animal intelligence. However, what sets us apart from animals is the fact that our Creator made us to have a special relationship with Him. And even when Adam and Eve placed *their* will above God's Word, He still loved us enough to pay the highest price to restore us to Himself. Jesus Christ lived in perfect obedience to God for us, and then suffered the penalty of our disobedience against God. In His Resurrection from the dead, all those who embrace Christ in faith receive the promise of being made new creations again—beginning right here in this life! *That's* the wide gulf between humans and animals!

Prayer: Dear Father, I thank You that You have given me being and life, and that when I was lost in sin, You still sought me out with Your gospel. Help me truly live as Your new creation in Christ. For His sake. Amen.

REF.: *Science Digest*, Aug. 1983. p. 73.

Beavers Build More Than Dams

Psalm 90:12

"So teach us to number our days, that we may apply our hearts unto wisdom."

Beavers are widely known for their huge constructions. However, beavers build more than just dams.

The beaver is quite an engineer. When building a dam, a beaver first rams poles into the bed of his waterway. The poles are placed on an angle to increase holding power, and are then rammed into the waterway bed, thick end first. This basic skeleton is then reinforced with twigs, leaves, mud and stone. It's not unusual for the beaver to use forked branches to further anchor his construction to trees and boulders.

One beaver dam was over 1,800 feet long, 9 feet high, and 18 feet wide. It was strong enough for a horse and rider to use as a natural bridge.

Large trees brought down for use in construction are not dragged by the beavers to their project. Instead, beavers dig overland channels and float the trees to the main body of water where they are working. These channels can be 1,000 feet long.

Our wise Creator gave so much intelligence and wisdom to the animal kingdom that often we can learn some things from the animals. Rather than supporting the idea of evolution, with its claim that there is progressively greater intelligence in higher creatures, the beaver supports the idea that the Creator is the source of all intelligence.

Prayer: Dear Lord, as the source of all intelligence, You have freely given intelligence to Your creatures in ways that confound those who would deny You. Grant me more wisdom than I now have and show me, through Your perfect Word, how to use it. Amen.

REF.: *Science Digest*, Aug. 1983. p. 73.

Construction Frogs

Job 38:36

"Who hath put wisdom in the inward parts? or who hath given understanding to the heart?"

Perhaps to get evolutionists to rethink their beliefs, the Creator has provided many inventive and unique structures. There are so many of these that even the longest evolutionary time scale could not account for their evolution.

Let's focus on a few examples among frogs for a moment. Frogs need to keep their eggs moist. Water must be available for their tadpoles. Frogs have many different ways of doing this.

The Venezuelan marsupial frog has a skin pouch, just like a kangaroo, except hers is on her back. The female marsupial frog stores her eggs in the pouch. There is plenty of moisture and protection in her pouch for the young frogs. When they are mature, the skin pouch sort of unzips and out come up to 20 young frogs!

The Brazilian tree frog uses a completely different approach. This frog actually builds a pool of mud, near a pond, in which it lays its eggs. This pool is usually built so that, by the time the eggs have developed into strong tadpoles, the regular tropical rains wash the pool, along with the swimming tadpoles, into the pond!

In Job 38:36 the Lord asks Job, "Who has put wisdom in the mind?" It is obvious to us that only the Lord, the Creator of heaven and earth, could have given frogs these abilities and this knowledge. And it is just as clear that it is nonsense to say that chance and time made these frogs and their abilities.

Prayer: Dear Father in heaven, I thank You that You are not only intimately involved in Your creation, but that You are also wiser than any created being. Help me to seek out Your wisdom in the pages of the Bible. In Jesus' Name. Amen.

REF.: Wilder, Rachel. 1983. Flying frogs. *Science Digest*, Nov. p. 120.

The Beetle Who Is a Chemist

Psalm 104:30

"Thou sendest forth thy spirit, they are created: and thou renewest the face of the earth."

Imagine a beetle that can move across the surface of water without any body movement. Some creatures have been given such ability and skill that it should be clear to even the most skeptical person that their Maker and Teacher is greater than any human being.

One such creature is the large-eyed rove beetle. Like many other insects, this beetle can walk on water. The surface tension of the water is enough to support the beetle on top of the water, given the special structure of his legs and feet. Should you see the large-eyed rove beetle, you will notice that he moves across the water, stops, turns, and starts up again without moving any part of his body. How does this work?

The large-eyed rove beetle has special glands that produce a chemical—something like detergent—that changes the surface tension of the water. When the surface tension of the water near his back feet is different from the surface tension of the water near his front feet, the beetle is pulled forward.

Such a unique method of movement requires sophisticated knowledge of both hydrodynamics and chemistry. Since the beetle is obviously not that smart, it is no wonder that evolutionists often speak of evolution as though it were intelligent and able to create such wonders. However, it makes a lot more sense to talk about a wise, personal Creator than it does to speak of the wisdom of a mindless force like evolution.

Prayer: Dear Lord, help me and other Christians to remember that as Creator of everything that science studies, You are able to see to it that Your Word is trustworthy in all matters it speaks about, even scientific matters. Amen.

REF.: *Science Digest*, Aug., 1983. p. 74.

The Bombardier Beetle

Genesis 1:24

"And God said, Let the earth bring forth the living creature after his kind, cattle, and creeping thing, and beast of the earth after his kind: and it was so."

The bombardier beetle has a powerful and complex system for protecting himself from enemies. Glands within the beetle produce a mixture of two hydroquinone compounds and hydrogen peroxide. These chemicals must be held in separate chambers, because when they come into contact with each other they react. When the beetle is threatened, the chemicals are mixed in a third chamber where a third chemical is added and an explosive reaction takes place. This reaction forces the caustic solution out through a special "nozzle." The beetle is skilled in aiming this nozzle at its enemy.

If this system is missing any of its parts, it is more than worthless to the beetle; the chemicals alone are very dangerous. So this sophisticated defense system could not have evolved in stages. Imagine the poor beetle who evolved the ability to produce the chemicals but hadn't yet evolved the right chambers to mix them without blowing itself up. The first time he became alarmed, *poof!*—no more beetle at all!

The bombardier beetle is such a problem for evolutionists that many of them have challenged claims that an explosion takes place. The fact is, the beetle can give you a nasty burn. Even the popular magazine *Science Digest* explained the bombardier beetle's defense with the words "explosive reaction."

Clearly the bombardier beetle is excellent evidence that creatures were originally created in finished form by a wise Creator, without the need for stages of improvement.

Prayer: I thank You, Lord, that You made a perfect creation that was so wonderful that even now, under the burden of sin, it still fills me with awe. However, I thank You with my life for saving me from sin, death and the devil. Amen.

REF.: *Science Digest*, Aug., 1983. p. 74.

The Venus Flytrap

Job 12:9-11

"Who knoweth not in all these that the hand of the LORD hath wrought this? In whose hand is the soul of every living thing, and the breath of all mankind. Doth not the ear try words? and the mouth taste his meat?"

The Venus flytrap is one of the most fascinating plants you will ever find. In college I had a Venus flytrap that I would now and again feed with a morsel of hamburger from the college lunchroom. One day a classmate saw me carefully wrapping up the morsel and asked what I was going to do with it. I explained that I was taking it back for my plant. He refused to believe that a plant would eat meat. Well, he stared in doubt as I placed the morsel of meat into one of the leaves and tripped the trigger. When the leaf closed on the meat, he fled the room in absolute disbelief.

Did you ever wonder how the Venus flytrap's leaf can close so rapidly on a fly or other morsel? It has taken scientists some time to figure out how it is done. When the trigger hairs are triggered, the cells on the outer layer of the leaf grow incredibly fast. The closing leaf is really the result of astonishingly rapid cell growth! It takes days for the cells on the inside of the leaf to catch up. As they do, the leaf slowly opens again.

The wild growth of the leaf's outer cells is started by an electrical impulse produced when something brushes up against the trigger cells inside the leaf.

The complex systems of the Venus flytrap are difficult to explain as the product of no intelligence. They are yet another example of the variations possible to an infinitely imaginative Creator.

Prayer: Father, there is no limit to Your imagination. Help me to understand that there is also no limit to Your love for me because of what your Son, Jesus Christ, has done for me. In His Name I pray. Amen.

REF.: How plants manage to move. *Science Digest*, Aug., 1983. p. 89.

Ants and Their Livestock

Exodus 4:11,12

"And the LORD said unto him, Who hath made man's mouth? or who maketh the dumb, or deaf, or the seeing, or the blind? have not I the LORD? Now therefore go, and I will be with thy mouth, and teach thee what thou shalt say."

Honeydew is one of the favorite foods of ants. There are a number of insects, including thorn bugs and aphids, who consume so much honeydew that they cannot digest it all. These insects are designed so that ants can milk them for the excess honeydew, just the way a farmer obtains milk from a cow.

Many species of ants will milk the honeydew-collecting insect by stroking its abdomen with its antennae. The tree hopper actually slows its secretion of excess honeydew to the milking ant so that the ant may drink his honeydew "fresh from the cow" so to speak.

Some ants construct clay pens and shelters in which they enclose certain species of aphids, just as a farmer keeps his herd. Like the farmer, the ants move the aphids to better honeydew feeding sites when necessary. They also care for the aphid eggs during the winter.

Herding ants are even better able to protect a young tree hopper than its parents. It is not unusual for a tree hopper parent to turn her young over to herding ants because they can do more for the young insect than she can.

Such complex relationships between creatures points to a God who is both Creator and Teacher—even to insects!

Prayer: Dear Father in heaven, as a good father, You try to teach me. I confess that I have been slow to learn and sometimes stubborn about learning. Forgive me for Jesus' sake and grant me Your Holy Spirit so that I may learn more from Your Word. Amen.

REF.: Thorn bug. *Science Digest*, Jan., 1985. p. 81.

Complex—But No Evolution

Psalm 46:10

"Be still, and know that I am God: I will be exalted among the heathen, I will be exalted in the earth."

According to evolutionists, coral is one of the earlier creatures on the evolutionary time line. If evolutionary theory were fact, one would expect the corals to be very simple. One would also expect them to have changed a lot in over half a billion years of life.

However, corals are not at all simple creatures. The stinger used by coral polyps to immobilize prey is very complex. It is produced by a highly specialized cell in the polyp itself. At one end of the stinging capsule is a poker, covered by a protective cap. At the other end of the capsule there is a hollow tube, coiled like a rope. The capsule itself is filled with a powerful poison.

When the poker end of the capsule is touched, a number of complex things happen in a very short time. The cap springs open, the poker rips a hole in the victim and continues in, carrying with it the poison tube, which then pumps the poison into the victim. All of these amazing steps take place in less than five ten-thousandths of a second!

Coral is "surprisingly" complex to be so near the bottom of the evolutionary ladder. In addition, the earliest corals in the fossil record—over half a billion years old by inflated evolutionary counting—are the same as those found today! So again, the scientific evidence fits well with the biblical story of creation.

Prayer: Lord, since You know all things, You cannot learn new things and improve the work You did during creation week. Even human sin has not been able to completely deface Your great genius. Assure me that because of Your great work for me on Calvary, my sin is forgiven and I shall be with You in eternity. Amen.

REF.: What's in a name. *Science Digest*, Feb., 1985. p. 87.

Why Birds Don't Need Socks

Job 12:7,9

"But now ask the beasts, and they shall teach thee; and the fowls of the air, and they shall tell thee: . . . Who knoweth not in all these that the hand of the LORD hath wrought this?"

Have you ever watched ducks out on an icy lake paddling around in the near-freezing water. They will spend much of the day in that one spot with the temperature well below freezing, and yet the cold doesn't seem to bother them. Perhaps on shore there are some sparrows hopping around in the snow. One wonders: why don't birds need shoes and socks?

If you or I ran around with bare feet and legs when the temperature was below freezing, it would not be very long before we had a good case of frostbite. We could even lose so much body heat that our very lives would be threatened.

Birds have a network of arteries that all blood going into their legs must enter. These arteries are interwoven with the veins returning from the foot. When the blood going to the foot enters this structure, called a "wonder net," its temperature is 106° F. The blood returning from the foot at the same time is only 37°F. The warm blood passing through the net reheats the cold blood coming from the foot before it enters the body. The result is that the bird loses very little body heat. So the blood going into the foot never becomes dangerously cold in normal winter weather.

The wisdom of this design is obvious. Could birds have survived having their feet frozen off before they learned to evolve this special arrangement? Our only conclusion can be that there is a Creator who cares for all of His creation, including you and me!

Prayer: I thank You, Lord, that Your hand is evident in all the creation. Help me to better see Your hand in my life, both leading me and protecting me. Amen.

REF.: Why don't ducks' feet freeze? *Science Digest*, Feb. 1984. p. 76.

What Was the First Form of Life?

Genesis 1:31

"And God saw every thing that he had made, and, behold, it was very good. And the evening and the morning were the sixth day."

Roman Vishniac is a renowned photographer of microscopic scenes. One of his favorite subjects is one of the so-called simplest forms of life. This "simple" life, the protozoan, is supposedly the ancestor of all living things. However, Vishniac has noted that the protozoan is neither simple nor archaic in character. This creature carries out all the functions of life that we require millions of cells to sustain—all within a single cell.

There is a parallel in technology. The first adding machines were large and heavy contraptions. As these mechanical adding machines became smaller, they had to be more cleverly designed *because* they were smaller. The first electronic calculators were even smaller and lighter than their mechanical brothers. But they were still much larger and simpler than calculators today. As calculators have decreased in size, they have had to be more cleverly designed because they are smaller.

Vishniac has learned that the so-called "simple" one-celled creatures are not simple at all. This is because they must carry out the business of life within the limits of their small size. This has not swayed him from evolution. However, this fact has convinced him that current theories are wrong. You see, evolution is in trouble if there are no "simple" forms of life. We believe that this clear evidence that there are no simple forms of life—evidence powerful enough to sway the experts—supports the Bible's claim that all forms of life were created in finished form by a wise Creator.

Prayer: Dear Lord, I come to You as the One who was the instrument of my creation and the only One through whom I have eternal life. For the sake of Your innocent suffering and death, let my earthly life be in You and my eternal life be with You. Amen.

REF.: *Science Digest*, Feb. 1984. p. 80.

An Intelligent Fungus?

Matthew 6:11
"Give us this day our daily bread."

Late at night, in hundreds of thousands of homes around the country, people are watching a bewildering array of monster movies. In these films, shadowy and not-so-shadowy monsters attack, possess or inhabit their victims. These stories are make-believe. However, in our real world, there is a fungus that does all these things.

These fungi grow in and on the bodies of insects. They begin their work when a single fungus spore lands on the body of an insect. The spore germinates by growing thread-like filaments through a weak spot or some other entry point in the insect's outer shell. Once inside the insect, these thread-like filaments grow into the insect's nervous system. Scientists have been amazed by the apparent fact that the fungus then seems to take over the insect, actually controlling its movements. It seems that the fungus directs its victim to climb a convenient plant to a height that will make the next generation of spores likely to land on another unsuspecting insect. It is there that the insect dies and the fungus produces its spores.

To the evolutionist, the ability of such a so-called "low" form of life to possess and control a more evolved form of life is totally without explanation. To those of us who believe in creation, intelligence is not a matter of evolution. Intelligence is the gift of a Creator. Only He has all intelligence and wisdom. There are no limits to His creativity as He provides His creatures with ways to proclaim Him as their Creator.

Prayer: Father, there are no obstacles for You as You fulfill Your promise to provide for all my needs. Help me to remember this as I faithfully look to You to meet my needs. In Jesus' Name. Amen.

REF.: Perry, Donald R. 1984. The creeping killers. *Science Digest*, Feb. p. 82.

Built-In Error Correction

Proverbs 3:11-12

"My son, despise not the chastening of the LORD; neither be weary of his correction: For whom the LORD loveth he correcteth; even as a father the son in whom he delighteth."

It seems like science fiction to suggest that if someone typed your name into a computer, misspelling your name, the computer would find it and correct it.

Yet the genetic code within each of your cells is an even more sophisticated information storage and transmission system. Not only does your genetic code store far more information in a microscopic space than our largest computers can, it has a built-in error correction system. Scientists have found a number of key enzymes in the cell that have just one job—find and correct errors in the genetic code. Those errors can creep in because of radiation, chemicals, or other reasons. These enzymes faithfully correct any errors, preventing mutations.

But these are the very ones scientists thought could cause evolution—before they discovered this process. One scientist who studied these enzymes said science has no explanation for how this process could have evolved naturally. Scientists also say they don't know how life could have continued without this genetic proofreading and correction.

Only our Creator could have been wise enough to design an information system that can correct errors within itself. Even we human beings have not figured out how to do this with our simplest computers. Asking us to believe that blind chance and natural law could have done it certainly strains the credibility of science!

Prayer: Dear Father, just as You have designed a chemical system to remedy errors in our genetics, Your plan of salvation by grace through faith in Jesus Christ's anointing work gives the full remedy for sin, including a new life. Grant me Your Holy Spirit so that I may be enabled to better live that new life You give me. In Jesus' Name. Amen.

REF: Fersht, A.R. 1980. *Trends in Biochemical Science*, v. 5. p. 282.
Lambert, G.R. 1984. *Journal of Theoretical Biology*, v. 107. p. 387.

Learning from the Dragonfly

Psalm 34:11

"Come, ye children, hearken unto me; I will teach you the fear of the LORD."

Evolutionary naturalism paints the picture of life as if it were a haphazard series of accidents that just barely resulted in a range of living things that just manage to survive. This fanciful picture destroys our sense of wonder over the sophisticated engineering designs in nature.

Consider the mystery of flight, for example. Some evolutionists suggest that perhaps birds are descended from lizards that fell out of trees a lot. Other evolutionists say that birds came from lizards who grew wings—not for flight, but to chase down and catch insects. Yet, they have little to say about the fact that we humans have come by most of our sophisticated knowledge about flight from studying the birds.

Then there is the problem, for the evolutionist, of how flight accidentally evolved *so many times* for so many creatures. Scientists studying the dragonfly are learning even more secrets of flight. Our best high-performance aircraft can barely lift themselves off the ground. However, the dragonfly can lift 15 times his own weight into the air. Scientists have learned that this is because the dragonfly's wings are designed to create little whirlwinds over their top surfaces. These whirlwinds are the secret to creating incredible lifting power. Ways are now being planned to apply this secret to new aircraft designs.

The engineering excellence found in nature and from which we have learned so much even in this day of interplanetary probes, is not witness to a mindless process of evolution, but to a wise and mindful Creator.

Prayer: Father, it seems a shame that humans learned only about flight from the dragonfly, and fail to see a witness to You. Yet, I know that I don't learn from You as I should, either. Forgive me for Jesus' sake and make me a better learner. Amen.

REF.: Dragonfly model for future wings *Science Digest*, Mar. 1984. p. 87.

Technological Secrets from God

Job 38:4

"Where wast thou when I laid the foundations of the earth? declare, if thou hast understanding."

A revolutionary design in metal-cutting blades is changing the metal-working industry. These blades don't cut like a knife. Instead, the blade is fixed in one position while the metal to be shaped spins on a lathe. This new-style blade stays sharp six times longer than the old blade design. And where titanium used to make the old-style cutting blades dull almost the second they were used, the new blades last up to 30 minutes.

This wonderful new blade was developed by two engineers who got the original idea from a rat. You see, rats' teeth always have a sharp edge. Engineers learned that this is because their teeth are hard on one side and soft on the other. So as they wear, they always keep a sharp edge. Using this principle for metal-cutting has lowered costs and increased productivity, since the blades don't have to be changed as often.

Beginning in Job 39, God humbles Job with several chapters of questions about the wise designs found in the animal kingdom. God's message is, "Job, if you think you're so smart, who created all of this with such wisdom?"

Even rats' teeth are created with such wisdom that, once we learned the principle, our technology was greatly improved. In view of this, it doesn't seem at all scientific to say that there is no Creator!

Prayer: Dear Father, help me not to think that I am wiser or smarter than I really am. Help me to seek Your wisdom and knowledge in Your Word, which is as high as heaven and as real as the creation. In Jesus' Name. Amen.

REF.: Self-sharpening blade has edge. *Science Digest*, Mar. 1984. p. 86.

The Fish that Goes Fishing

Exodus 34:7a

". . . keeping mercy for thousands, forgiving iniquity and transgression and sin. . ."

A family of fish that live in the depths of the sea actually go fishing—pole, dangling bait, and all. This is only natural, since they, like people who go fishing, like to eat fish.

This group of fish is collectively called the angler fish. The front dorsal spine in the female angler is located on her head. It is much longer than any of the other dorsal fins and has a fleshy bait on the end. When hungry, she sits very still and dangles the bait in front of her mouth. It doesn't take long before something swims up to investigate this bait as its possible lunch. But before it can think about taking a sample, the angler has turned the tables and eaten her own lunch.

Of course, this system wouldn't work very well in the pitch darkness one mile below the surface of the ocean without one other special provision. Like many other fish at such depths, the angler is able to generate her own light. In her case, only the bait lights up, so she stays invisible to her victims. The chemical method she uses to create the light is nearly 100% efficient!

In His imaginative creativity, God was pleased to place creatures a mile below the surface of the sea. Yet, in His mercy, He has provided them with everything they need to make their living there. Likewise, you may not know why God has placed you into a particular situation right now. If He so cares for the creatures a mile below the waves, how much more does He care for You. And He has even included a light for you in His Son, Jesus Christ!

Prayer: Father in heaven, I can see from what You have created that You are kind and caring. I learn from Your Word that You are also gracious and forgiving for the sake of Your Son, Jesus Christ. I ask that this truth would always be in my sight. Amen.

The Truth in Silver

John 5:46

"For had ye believed Moses, ye would have believed me: for he wrote of me."

The Bible itself tells us that the first five books of the Bible were written by Moses. Jesus Himself calls Moses the author of the first five books of the Bible over a dozen times. Unfortunately, liberal biblical scholarship has rejected these claims. These scholars have tried to teach Christians that the first five books of the Bible were written over many generations and finally finished only five centuries before Christ's birth.

Now an amazing new discovery is forcing these scholars to change their teaching and has helped to scientifically confirm what Jesus said. Archaeologists have discovered two tiny silver scrolls on which are carved important passages from Moses. The scrolls, once used as amulets, were found in the ruins of early sixth or seventh century B.C. Jerusalem.

Each scroll was made of thin silver foil, no thicker than tissue paper. One of the scrolls was an inch wide and an inch and one-half long. Scientists studying the fine engraving under a microscope found what appears to be the text of a priestly blessing from the Book of Numbers. They noted that it was remarkable that the blessing is identical to modern texts, showing that the text has been accurately preserved longer than they ever imagined.

True faith doesn't require the support of science. However, we are right to expect that true science will never do less than support the Bible when they are both speaking about the same subject.

Prayer: It grieves me, Lord, that so few today understand that the truth of Scripture is important because all of Scripture speaks of You. Help me to remember this so that I may make right and profitable use of Your Word. Amen.

REF.: A startling revelation about the Good Book. *Discover*, Aug. 1986. p. 10.

A Six-Foot Bat

1 Corinthians 2:9-10

"But as it is written, Eye has not seen, nor ear heard, neither have entered into the heart of man, the things which God hath prepared for them that love him. But God hath revealed them unto us by his Spirit: for the Spirit searches all things, yea, the deep things of God."

Could the giant flying fox be a primate? In the 18th century, scientists weren't quite sure what to do with bats. Some wanted to classify them as primates, while others thought they should have their own category. Since 1780, all 950 known species of bats have been classified in their own category. The bat order is divided into microbats and megabats. Microbats have an average wingspan of less than a foot. Megabats are larger. The giant flying fox has a wingspan of six feet!

The evolutionary link between microbats and megabats is dissolving as scientists learn more about them. Although megabats and microbats are very clearly bats, they are constructed in importantly different ways. Microbats rely mainly on echo-location to navigate. Megabats have very large eyes, so that they can see like owls in low light. The nerves that connect the eyes of microbats to the brain cross. As a result, the image from the right eye goes to the left brain and the left eye feeds its image to the right brain. However, megabats are wired with both eyes feeding into both sides of the brain. This arrangement is otherwise unique to *primates*!

If bats evolved, these new findings mean that bats must have evolved at least twice! It also means that some bats evolved from monkey-like creatures!

It makes far more sense to admire God's unlimited creativity. We can logically reason that God gave His larger bats eyesight like that of a monkey, because their needs are similar.

Prayer: Dear Father, I thank You that what You have made is so pleasing to the mind and to the eye. I ask that our growing scientific knowledge would continue to make it more difficult to deny You. In Jesus' Name. Amen.

REF.: The eyes have it: megabats are primates. *Discover*, June 1986. p. 7.

Mastodon Lunch

Joshua 24:15b

"...but as for me and my house, we will serve the LORD."

Mastodons were distant relatives of the elephant. Modern science says that one of the main parts of their diet was conifer needles. Supposedly they died out at the end of the Ice Age because conifers became scarce as the world warmed.

Now the discovery of the last meal of one mastodon calls this theory into question. Living bacteria have been found in a mastodon that died during the Ice Age. The bugs were found in the remains of a mastodon's last meal, preserved with a skeleton that was discovered while preparing a golf course in Ohio. Scientists think the bacteria normally lived inside the mastodon, working as part of the mastodon's digestive process.

What scientists discovered about the mastodons' diet could upset some theories about what drove them to extinction. The scientists said the four-ton mastodon had spent its last day eating a salad of swamp grass, moss and leaves, a bit of seed, and a garnish of water lily. What was left of these plants, as well as the bacteria, were separated from a mass of reddish-brown matter found near the mastodon's ribs and vertebrae.

Scientists who believe in evolution claim that the animal died 11,000 years ago. But how sure can we be about that? After all, evolutionary scientists were wrong about the mastodon's diet. As a result, their theories about why mastodons became extinct seem to be wrong. The facts that the bacteria were still alive and the stomach contents were not fully decayed tells us that this evolutionary date, too, is greatly exaggerated.

Prayer: Dear Lord, so many people prefer to reject Your Word and then do their own speculating about things You have clearly revealed to us. I ask that through Your Holy Spirit, You would help me not to speculate where Your Word has spoken so clearly. Amen.

Making Scents of Smells

Genesis 27:27

"And he came near, and kissed him: and he smelled the smell of his raiment, and blessed him, and said, See, the smell of my son is as the smell of a field which the LORD hath blessed."

How many different scents do you think your nose is able to smell? The answer is about 10,000. How does your nose tell the difference between a rose and a skunk? Until only recently, scientists really weren't sure.

Now we know that humans and mammals detect smells by using at least 1,000 different genes. These genes are located in every cell in your body, but they are active only in your nose. This discovery, say scientists, gives them their first insight into how smell works. Before this, said one scientist, "We had almost no knowledge of the olfactory system and this really opens up the molecular study of smell in a very fascinating way."

They say that each of the different genes seems designed to respond to a small group of odors. As each group of genes does its part, the entire system is able to recognize everything we can smell. These genes pass signals from the nerve cells in the nose to a small region of the brain called the olfactory bulb. Now that smell genes are identified, researchers say they can begin to understand how odors are detected and how the brain interprets those signals.

Our sense of smell relies on a very elegant and sophisticated system. Imagine, 1,000 genes lie at the heart of that system. Each one of those genes is information—a programmed code that is designed to make sense of what we smell. Information, programming and decoding are all terms that assume that we have a highly intelligent Creator who designed our sense of smell!

Prayer: Dear Father in heaven, I thank You for my ability to detect scents. As scientists continue to learn how You have designed our sense of smell, cause them to recognize Your genius and so be drawn to the cross of Jesus Christ. Amen.

REF.: Scientists find genes that let nose make scents of what it smells. Minneapolis *Star-Tribune*. Apr. 5, 1991. p. 7A.

Winged Troops

Acts 10:7

"And when the angel which spake unto Cornelius had departed, he called two of his household servants, and a devout soldier of them that waited on him continually;"

We have long known that the community within a honeybee hive divides the labor among specialized workers. There are scouts who search for food. Nurses tend the young. Some bees specialize in cleaning. There are even undertaker bees who remove and dispose of dead members of the hive.

Scientists have now discovered that honeybee colonies also have brigades of specialized "soldiers." These bees often seem to be sitting around, not doing much. In fact, they have only one duty—attack large intruders. It appears that these bees are fed and cared for by the hive members while they wait for the alarm. Should a raccoon, bear or human threaten the hive, thousands of soldier bees swarm around the enemy and sting.

Soldiers have been well-known members of ant and termite colonies. However, up to now, they have never been confirmed in honeybee hives. The universal need for soldiers for protection among social insects reminds us that some things are universal in this world. These soldiers are not responsible for a bear's attack on the hive. Nor are human soldiers who follow just war principles responsible for human war.

Humanity's greed and lack of love for others cannot be abolished by getting rid of armies. Rather, greed, theft and the need for protection remind us that we are not what God intended us to be. The solution lies not in denying reality, but in returning to our Creator through the forgiveness of our sins as made possible by Jesus Christ.

Prayer: Father, I must confess that even as I love You and agree that Your will is good, I still sin. Forgive me for the sake of the innocent suffering and death of Your Son, Jesus Christ. May His goodness, and not my own, always be my assurance of Your love. Amen.

REF: Soldiers swarm to rescue. *The Argus* (Fremont, CA), Mar. 17, 1991.

Canaanite Golden Calf Discovered

2 Kings 10:29

"Howbeit from the sins of Jeroboam the son of Nebat, who made Israel to sin, Jehu departed not from after them, to wit, the golden calves that were in Bethel, and that were in Dan."

The Old Testament mentions the worship of a "golden calf" several times. According to the Bible, the Israelites fashioned one of these idols after their exodus from Egypt. Later, king Jeroboam worshiped the golden calf idol.

Until now, no one in our time had ever seen one. In the summer of 1991, Harvard University archaeologists excavating Canaanite ruins near the ancient port city of Ashkelon unearthed a "golden calf." Ashkelon is in Israel.

This "golden calf," however, is not made entirely of gold. It is about four and one-half inches long and four and one-fourth inches tall. It weighs just under a pound. The arms, legs, horns, tail and other parts were made of different metals and were attached to the body in sockets. The body itself was bronze. The calf was dated through the pottery and other artifacts found alongside it to about 1550 BC. This would mean that the calf is Canaanite.

Though there was no reason to question what the Bible says about golden calf worship, many scholars preferred to refer to biblical statements on the subject as legends. This discovery shows the Bible's accuracy when it speaks of golden calf worship. It also shows that the Canaanites worshiped these idols before the Israelites took over the Promised Land. We should note that no archaeological discovery in the Holy Land has ever shown the Bible to be anything less than accurate. So-called experts, who have been skeptical of the Bible's claims, have been proven wrong many times.

Prayer: Lord, I thank You that Your Word is truth. I, too, live in a world with many idols. I ask that You would help me see how the world's idols influence my life. Help me to be faithful to You, both in my life and witness. Amen.

REF: Before-Christ 'golden calf' idol startles archaeologists in Israel. *The Argus* (Fremont, CA), July 25, 1990.

Robot Bee Communicator

John 1:10

"He was in the world, and the world was made by him, and the world knew him not."

Scientists have, for the first time, successfully communicated with honeybees in their own language! Scientists have for some time known how to interpret many of the elements of the bees' so-called "waggle dance." This dance is used by scouts to communicate the location of food sources to the food gatherers in the hive.

A couple years ago, researchers decided to see if they could confirm their understanding of bee communication. But how do you talk to a bee and get confirmation that she understood what you said? Scientists decided that the only way to do this was to build a robot bee through which they could communicate. The first attempts at building a robot bee acceptable to real bees did not work well. Early models were viciously attacked.

Scientists finally perfected a computer-controlled robot bee that real honeybees accepted. These scientists describe honeybee language as elegant and precise. Yet they found they had learned enough bee language to communicate successfully a food source and its location to real bees. They also discovered that they have much more to learn about bee language. As a result, it may someday be possible for farmers to direct bees to crops that need pollination.

Just as bees can best understand other bees, we humans can best understand other humans. That explains why, when God our Creator wished to communicate with us, He became a human being. That human, Jesus Christ, taught us about God; and He also made peace with God for us, making possible the forgiveness of our sins.

Prayer: I thank You, Lord, that You became a human being so that I might have my sins forgiven. I also thank You that in becoming human, You helped us understand our Creator's loving intentions toward us. Help me to communicate this to others. Amen.

REF.: Weiss, Rick. 1989. New dancer in the hive. *Science News*, v. 136, Oct. p. 282.

The Heavens Declare the Glory of God

Genesis 1:16

"And God made two great lights; the greater light to rule the day, and the lesser light to rule the night: he made the stars also."

Astronomers who believe in evolution tell us that the stars formed when gas in a nebula began to gather due to gravity. Eventually, that gravity became strong enough to collapse the gas into a star.

Astronomers who believe in creation point out that this is unlikely, if not impossible. As we all know, compressed gas wants to expand by moving to areas of lower pressure. A cloud of gas that was large enough to begin compressing under gravity would begin to rotate faster. Magnetic forces would also increase. The increase in pressure, magnetic forces and centrifugal forces would overcome the forces of gravity, preventing the formation of a star.

At the same time, blue stars are common in the arms of the galaxies we can see. These stars are considered young because they burn their fuel so fast that they could only last a few million evolutionary years. On the evolutionary scale, they must have formed long after the Big Bang. The number of blue stars would then indicate to an evolutionist that star formation is common. Yet, as we have seen, the forces at work to form a star in this way make these stars either unlikely or younger than millions of years.

Perhaps God created the blue stars to serve as a clear indicator to scientists that He created the stars fairly recently.

Prayer: I thank You, Father, for the beauty of the created heavens. The very existence of the stars praises Your Name. Amen.

Ref: *Answers Magazine*, 10-12/07, pp. 78-80, Jason Lisle, "The Stars of Heaven Confirm Biblical Creation."

Self-Centered Ants

Philippians 2:3

"Let nothing be done through strife or vainglory, but in lowliness of mind let each esteem other better than themselves."

It's a challenge, even for the Christian, to get along with lazy, selfish people. This is because they also tend to be quarrelsome, especially when their desires are not met. Now scientists have discovered that the phrase "industrious as an ant" does not apply to all ants. They report that they have found an entire ant society that is lazy, selfish and quarrelsome.

Sometimes a society is pleasant to live in and sometimes it's not. Not all ant societies are harmonious examples of cooperation. One ant society, popularly called the slave-makers, must be a miserable place to live. Slave-making ants make their living by plundering the nests of other species and making slaves of their captives. Slave ants are forced to feed, house, and even defend their masters.

This leaves the slave-making ants with a great deal of idle time. Most of that idle time is spent in endless squabbles with each other over social status. The higher a slave-maker's status, the more service from a slave she receives. It would seem that just as idle human hands can lead to strife between people, idle ants tend to fight with each other too.

This universal linking of selfishness, laziness and fighting tells us that there are indeed universal moral truths. It shows us that those who say there is no absolute morality are wrong. Perhaps God has allowed this ant society to corrupt itself so we could have a scientific basis to reject today's moral relativism.

Prayer: I thank You, Father, that You wish me to live my life in joy rather than strife. Forgive my selfishness and tendency to feel that I have not received as much as I deserve. Fill me, instead, with thanksgiving toward You and love for those around me. In Jesus' Name. Amen.

REF.: Antagonistic ants. *Discover*, Nov. 1983. p. 14.

Life that Can't Exist

Ecclesiastes 8:16,17b

"When I applied mine heart to know wisdom, and to see the business that is done upon earth: (for also there is that neither day nor night seeth sleep to his eyes:). . . though a man labour to seek it out, yet he shall not find it; yea farther, though a wise man think to know it, yet shall he not be able to find it."

The founders of modern science, on the whole, believed in the Creator. The methods they developed were based upon the truth that the creation is voluntary. This means that God was not confined to what we have later defined as "scientific laws." God could make anything He wanted in any way He wanted. Science cannot be done from the philosopher's easy chair. Scientists must go out into the field and study how nature works. When science follows this path, it is open to astonishing discoveries.

One recent unexpected discovery was the hot water vents deep below the ocean's surface. Where the earth's crust has split apart, hot water gushes out. This water reaches temperatures hot enough to, say, cause a piece of paper to burst into flame. Only the tremendous pressure more than a mile below the waves keeps the water in liquid form.

It was in this environment that scientists made an even more amazing discovery. They found bacteria that could live in the almost 700° water. These amazing creatures find the normal, sea level boiling point of water to be so *cold,* they can barely survive. Scientists are still trying to learn how these creatures live at temperatures that would destroy the normal biochemistry of life.

Each new discovery tells us a little more about the work of our wonderful Creator. That's what science was originally intended to do. Keep that in mind when you hear someone say that belief in the Creator makes doing science impossible.

Prayer: Dear Lord, how little we have learned about Your wonderful work all around us! I ask that You would give extra boldness, strength and wisdom to Christians who are working in science, especially when they are attacked for believing in You. Amen.

REF.: Thomas, Lewis. 1983. On life in a hell of a place. *Discover*, Oct. p. 42.

Flying Spiders

Romans 11:33

"O the depth of the riches both of the wisdom and knowledge of God! how unsearchable are his judgments, and his ways past finding out!"

Spider silk begins as a liquid protein made by silk glands on the spider's abdomen. Spiders make many kinds of silk for different uses. As the liquid silk is forced through the spider's spinnerets, it begins to dry. The spinnerets pull and stretch the silk, creating just the right kind of silk for the spider's use. Though the result seems thin and weak to us, ounce for ounce, spider silk is stronger than steel.

Web-building spiders make two types of silk for their traps. The basic structure of the web is made of strong, nonsticky silk. Then the spider adds a sticky, elastic silk to trap its prey. Some webs are irregular, others are flat and sheet-like, while still others are shaped like funnels. All of us have admired the beautiful orb web.

However, spiders also have many other ways in which they use their silk. The ogre-faced spider makes a net that it throws around its intended victim. The European water spider builds its web underwater. It stores its air supply in the web. The lasso spider twirls a single strand of silk over its head. A tiny drop of sticky silk at the end of the lasso captures its prey when the lasso is thrown. Probably the most creative use of silk is by the young of some species. When they hatch, they produce long loops of silk that catch the wind and send them floating off to new places.

The spider's skill in producing and using silk had to come from Someone wiser and more clever than the spider. Clearly the spider has been well-equipped and instructed by its Creator.

Prayer: Father, I thank You for the intelligence and wisdom You have so freely given to Your creatures. Help me to treasure, most of all, Your wisdom in the work of Jesus Christ for my salvation so that my life, too, may glorify You. In Jesus' Name. Amen.

REF.: Pinkston, William S., Jr. 1980. *BIOLOGY for Christian Schools*. Greenville, SC: Bob Jones University Press. p. 413.

The Most Famous Hurrian

Genesis 12:4b

"And Abram was seventy and five years old when he departed out of Haran."

Excavations north of Baghdad, in Iraq, are helping us to understand one of the most famous Old Testament characters. The more than 4,000 ancient documents that have been found there shed detailed light on the life and times of Abraham. Abraham was from Haran, which was an important city in the Hurrian empire.

Scholars are finding that many of the sometimes puzzling activities of the patriarchs make sense in view of Hurrian laws and customs. For example, you will remember that Abraham twice tried to pass off Sarah as his sister. This was evidently his rather strange attempt to avoid being murdered by someone who might want her. Isaac later tried the same ruse. Hurrian documents reveal that they were simply invoking the Hurrian custom of wife-sistership. This was a legal arrangement one step above wife status. Under Hurrian law Sarah was probably Abraham's sister in this sense. This explains why the same strategy keeps turning up.

Another occurrence that seems strange to us took place when Esau sold his birthright to Jacob for a pot of stew. However, Hurrian documents show that it was common to sell one's birthright, often for items of small value. When Sarah was unable to have a child and offered her handmaiden to Abraham, she, too, was only following Hurrian law.

As we learn more about the world of the Bible, we find that the Bible's integrity remains solid. That doesn't surprise those who accept the inerrancy and authority of Scripture.

Prayer: I thank You, Lord, that You have revealed Your Word to us and have preserved it over thousands of years. Forgive me for those days when I thought I was too busy to study Your Word, and fill me with new eagerness to learn more. Amen.

REF.: West, Stuart A. 1981. The Nuzi tablets. *Bible and Spade*, v. 10, n. 3-4. p. 65.

Designed for Flight

Job 39:26

"Doth the hawk fly by thy wisdom, and stretch her wings toward the south?"

Many textbooks tell young people today that birds are modified reptiles. Suppose, they say, that millions of years ago the scales on some reptiles began to fray along the edges. In time, they say, the frayed scales turned into feathers and birds were born.

The elegance and beauty of the feather make this story hard to believe. Can sticking a feather on a lizard produce a peacock? The bird's feather is only a small part of the complete flying system of the bird. Even with very careful planning and redesigning, a reptile doesn't have what it takes.

A bird needs massive breast muscles for flight. In some birds, 30% of the body weight of the bird is breast muscle. By comparison, in humans, breast muscles are only about 1% of body weight. A bird also needs an extremely high metabolism and blood pressure to deliver the energy those muscles need for flight. Birds have a higher metabolism than any other creature; they also have the necessary high blood pressure. Finally, as is well known, birds need light skeletons. The man-o'-war has a wingspan of seven feet. But its entire skeleton weighs only a few ounces—less than its feathers!

Even the most clever rebuilding of a reptile cannot produce a bird. In fact, birds have very little in common with reptiles. The entire being of the bird, from body to brain, has been specially designed for flight by a Creator who clearly knows everything there is to know about flight.

Prayer: Dear Father in heaven, the beauty, grace and huge variety of birds You have created are some of the most beautiful creatures on earth. As I thank You for them, fill me with new wonder over the wonderful work of Your hands. Amen.

REF.: Vandeman, George. 1991. The miracle of flight. *Signs of the Times*, May. p. 25.

Greenhouse Thermostat Discovered

Jeremiah 10:2

"Thus saith the LORD, Learn not the way of the heathen, and be not dismayed at the signs of heaven; for the heathen are dismayed at them."

The world is a frightening place without God. This fear can be seen today in fears that modern people often confuse with science.

Several years ago stories of gloom and doom about a greenhouse heating of the earth were starting to fill the media. We at Creation Moments looked at the scientific basis for these claims. We wanted to know if the greenhouse warming of the earth was a true threat or just another fear of a godless people. We then published the conclusion that the thin and conflicting evidence for greenhouse warming most likely came from the heat island effect of growing cities around some of the temperature measuring sites.

Knowing the elegance with which God designed everything in creation, creationists suspected that God probably designed the earth with built-in thermostats that help keep it from warming up too much. When the Pacific Ocean warms more than normal, it sends increasing amounts of water vapor into the air. If heating continues, the water vapor will actually cause the heating to increase until vapor is forced higher into the atmosphere, creating icy cirrus clouds. These clouds reflect enough incoming sunlight to shut down the temperature increase and return things to normal.

We are called to be good stewards of the earth. We can be good stewards in joyful service, freed from the fears that afflict those who don't know God.

Prayer: Dear Father, it is very easy for us to pick up the fears of those around us who don't know Your personal love and care. Forgive me when I, too, have shared the fears of the unbelieving world and strengthen my faith. In Jesus' Name. Amen.

REF.: Clouds keep ocean temperatures down. *Science News*, v. 139, May 11, 1991. p. 303.

Medical Light on Gene Research

Mark 2:17

"When Jesus heard it, he saith unto them, They that are whole have no need of the physician, but they that are sick: I did not come to call the righteous, but sinners to repentance."

Fireflies may help light the way to new medical breakthroughs. The gentle glow of a firefly on a summer evening is produced by the chemical *luciferase*. The firefly has a gene with the code that makes this chemical. When that gene is added to the genetic information of other plants or animals, the cells begin to glow harmlessly. Researchers have been using the firefly gene to test their ability to add genetic information to a cell.

When human heart arteries become clogged, a medicine called TPA is among the treatments doctors use. TPA is a natural clot-dissolving protein that helps prevent heart attacks. The problem is, without a continuous supply of TPA, the artery may reclog. It would help if the artery could make its own TPA 24 hours a day.

Since TPA is a natural protein for which there is a genetic code, researchers wanted to know if genetic information could be added to artery cells. To find out, they injected dogs with the genetic code for luciferase. If genetic information could be added to the cells through this method, they would actually see the results in the glowing tissue. Tissue samples taken three days later did, in fact, have the distinctive glow of the firefly proving that the method works. They warn that several years of research lie ahead before artery cells may be able to dissolve their own clots.

Those glowing fireflies that delight children and adults show God's design in the creation in an unexpected way. The chemical that makes the glow is proving to be an important tool of medical science.

Prayer: Lord, I thank You for the blessings of modern medicine. I pray that You would bless researchers who are working to make us healthier. Let the beauty of Your handiwork remind them that they have a Creator to Whom they are accountable. Amen.

REF.: Glowing evidence of gene-altered arteries. *Science News*, v. 139, June 22, 1991. p. 391.

Moths Poison Eggs

Genesis 45:5

"Now therefore be not grieved, nor be angry with yourselves, that ye sold me hither: for God did send me before you to preserve life."

In a clever scheme of self-protection, the more poisonous the male rattlebox moth is, the more prized he is by the female rattlebox moth.

In the caterpillar stage, the male moth favors the pods of certain legumes. These pods contain a powerful poison that seems to have no effect on the caterpillar. After the caterpillar has become a moth and it's time for mating, the male releases a scent produced by the poison. Those who have more poison in their bodies release a stronger scent, while those with little poison release only a weak scent. Female rattlebox moths prefer males with a strong scent. During mating, the male coats the eggs with some of the poison, making them unpalatable to predators, like ladybugs, who might eat the eggs. So the more poison a male has, the better protected the female's eggs will be. Studies have shown that female rattlebox moths will ignore males that have collected no poison.

Likewise, the male of the queen butterfly species collects another plant poison in his system. This poison is produced by plants for protection against insects. The butterfly doesn't attack the plant; rather he drinks the poison from already injured plants. Here, too, the female queen butterfly favors those males with the most poison.

It appears that human beings are not the only creatures who go to the drug store to get needed medicines. The Creator has provided the animal world with their own drug store and the knowledge of how to use the chemicals He has provided.

Prayer: Dear Lord, as I see Your intelligence and individual care at every level of the creation, I glorify You. Help me to remember how loving and clever You are in all things, especially when I don't see a solution to some problem in my life. Amen.

REF.: Different strokes for six-legged folks. *Discover*, June 1988. p. 12.

Baby Immunizes Mother

Galatians 4:6

"And because ye are sons, God hath sent forth the Spirit of his Son into your hearts, crying, Abba, Father."

Most of us are familiar with the body's response to foreign tissue. Transplant patients must be carefully matched and then patients still must be given powerful drugs to prevent destruction of the transplanted tissue. Our bodies' immune system identifies tissue that's not its own and then tries to destroy that tissue.

As we have learned more about the immune system, doctors began asking themselves why a pregnant mother's body doesn't reject the foreign tissue of the baby growing inside her. The answer to this mystery is not fully understood, but research has suggested some answers.

From conception, the unborn child is a genetically unique individual. Researchers think that the most important protection the unborn child has is the *trophoblast.* This tissue, known as the caul, seals the growing infant from its mother. However, the trophoblast itself develops from the same cells as the infant. So why doesn't the mother's immune system attack? It appears that under normal conditions the trophoblast is a special set of cells that the body cannot identify as foreign. In addition, the trophoblast makes cells that encourage the mother's immune system to protect it. Scientists now believe that this amazing feature may actually be provided by the *father's* genetic information.

In effect, the baby, with the help of the father, immunizes the mother's system so that it doesn't reject the baby! So, even before birth, the mother nurtures and the father protects the child.

Prayer: Father in heaven, I marvel at the wondrous system You have designed for reproduction. Always keep me under Your protection, and help me to grow to maturity in this life as I look forward to eternity with You. In Jesus' Name. Amen.

REF.: Wechsler, Rob. 1988. Hostile womb. *Discover*, Mar. p. 83.

Listening to Plants

Genesis 2:10

"And a river went out of Eden to water the garden; and from thence it was parted, and became into four heads."

Some people have the green thumb in their family. Others may complain that they fail with plants because plants don't *tell* them when they need care. Of course, most people can do well with family pets, who are enthusiastic about letting their needs be known.

Now scientists have learned that plants do indeed cry out when they need attention. We just haven't heard them because the sounds made by a thirsty plant are about five times higher-pitched than we can hear. But scientists with tiny, specialized microphones have heard corn plants calling for water. During their study, researchers learned that plants make a great deal of noise. Corn leaves and stalks make noises as they slide against each other during growth. We have all heard the sound of plant leaves rattling in the wind. Corn stalks also make noise as they bend in the wind.

However, another sound was heard when the corn didn't have enough water. It was a high-pitched popping noise. Plants have water tubes in them that carry water and nutrients up from the roots to the leaves. Water flows up these tubes under tension. When there is not enough water, the tension becomes too great and the tubes fracture. The popping is the sound of those tubes fracturing. The result of fractured tubes is a wilted plant. When water is restored, the tubes refill, the popping noises stop, and the plant again looks healthy.

The purpose of this research was to learn when to water crops and how much water they need. Now that we have learned how plants communicate, crops can tell farmers when to water!

Prayer: I thank You, Lord, for the generous supply of food with which You have blessed us. Prosper the work of those who are trying to learn how to get the most food with the fewest resources so that those most in need may have food, too. Amen.

REF: Scientists say plants may tell of watering needs in future. *Wichita Eagle-Beacon*, Jan. 12, 1986.

Scientists Question Age of Universe

Psalm 8:3-4

"When I consider thy heavens, the work of thy fingers, the moon and the stars, which thou hast ordained; What is man, that thou art mindful of him? and the son of man, that thou visitest him?"

If your measuring stick is inaccurate, you could end up with too much wallpaper or too little. If your map tells you that your destination is 200 miles away when it's really only 50 miles away, you might drive right past it. If your map of the universe is inaccurate and stars are a lot closer than you expect, you will arrive at the wrong age for the universe.

Most science textbooks solemnly teach that the universe is 10 to 20 billion years old. A distant galaxy in the Virgo cluster is one of the primary measuring sticks used by astronomers to figure out the distance of other objects in space. Ultimately, these distances are used to compute the age of the universe. Recently, though, astronomers were finally able to see individual stars in this distant cluster.

Using a new, high-resolution camera, astronomers saw stars that are very much like stars in our own Milky Way galaxy. Since they know how bright these stars are, they were able to check their computations on how far away the measuring-stick galaxy is. Instead of the expected 100 million light-years, their new figures told them that the galaxy was only 50 million light-years away. This means that the universe might only be half as old as they thought.

The lesson for us in this is that the millions and billions of years many scientists talk about are mere speculation. Many measuring sticks for the age of the universe are available, and many of them support an age in the biblical range of only thousands of years.

Prayer: Father, the wondrous sky above our heads clearly shows forth Your power and might. I thank You for its beauty. Help me not to be fooled or intimidated by the claims of unbelieving scientists. In Jesus' Name. Amen.

REF.: Virgo images suggest smaller universe. *Science News*, v. 139, June 15, 1991. p. 381.

Those Amazing Ants

Matthew 6:25

"Therefore I say unto you, Take no thought for your life, what ye shall eat, or what ye shall drink; nor yet for your body, what ye shall put on. Is not the life more than meat, and the body than raiment?"

How many ants are there in the world? No one has actually counted them all or asked them to fill out little ant census forms. However, scientists have estimated that there are about *10 million billion ants* in the world, waiting for you to have a picnic. That's more ants than there are mammals, reptiles, birds, and amphibians combined.

God made so many ants because ants are important housekeepers for the earth. Ants, not earthworms, turn most of the world's soil, drain it, and enrich it. Ants dispose of 90% of the corpses of small dead animals. As the world's gardeners, ants spread and plant more seeds than any other creature.

Like bees, ants have a sophisticated society, which includes workers, nurses, soldiers, hunters, farmers, and even builders. Each of these different classes have specialized organs for their work. The soldier ants of one species are walking bombs. They are loaded with poison and literally explode when under attack, spraying poison on their enemies. Army ants in Central and South America march across the countryside in a line that can be nearly 50 feet long. The solid column sweeps along, flushing out even small animals, as it looks for food.

The huge variety of living things around us represents more than simply variety. That variety carefully includes creatures to do every job that needs to be done to maintain life on earth. This, in itself, is a witness to both the creativity and wisdom of the Creator.

Prayer: Lord, I thank You for the way You have taken care of the needs of the earth. If it had been up to us, we would have forgotten some needs and not known about others. So I place my past, present and future in Your loving hands. In Jesus' Name. Amen.

REF.: Wilson, Edward O. 1990. Stalking the mighty ant. *Discover*, March.

A Flood of Flood Legends

2 Peter 3:5-6

"For this they willingly are ignorant of, that by the word of God the heavens were of old, and the earth standing out of the water and in the water: Whereby the world that then was, being overflowed with water, perished:"

Today it is considered scholarly to reject the Bible's account of a worldwide flood. Some try to compromise by saying that the flood recorded in the Bible was only a local event. The problem with these scholarly claims is that there were too many *witnesses* who disagree with them.

We can test these scholars' claims. If the Bible's account of a worldwide flood is true, it was witnessed by every person on earth. The story of this event would have been passed down to their descendants and spread across the whole face of the earth.

Researchers have catalogued some 270 stories of an ancient destructive flood in various cultures around the world. A large majority of these stories have been shown to predate any Christian influence. As one would expect, details differ after such a long time. What's remarkable, though, is that where the details of these stories agree with the Bible, they also tend to agree with each other. For example, the ancient Greek flood hero was told to build and stock an ark because the god Zeus wished to destroy humanity. The Aztec story of a universal flood says it took place 1,716 years after the creation of the world—almost the same as the Bible's date! The Babylonian flood story shares seven major details with the Bible's account. Similar stories are found around the world, including Australia, India, Scandinavia and China.

The universality of the flood stories and their similarity to the Bible cannot be explained unless they are based on an event that actually took place.

Prayer: Dear Father in heaven, I thank You that Your Word is sure and true and that it has been accurately preserved to our day. Help me to live Your Word in my life and tell others about the wonderful things You have done. In Jesus' Name. Amen.

REF.: Roth, Ariel A. 1990. Flood stories—can they be ignored? *Origins*, v. 17, n. 2. p. 51.

The Universal Boomerang

Acts 17:26

"And hath made of one blood all nations of men for to dwell on all the face of the earth, and hath determined the times before appointed, and the bounds of their habitation;"

When boomerangs are mentioned, most people think of Australia. Boomerangs have been in use there for thousands of years. Historians have always been puzzled about how ancient humans invented the boomerang.

The boomerang is designed to take advantage of some very sophisticated aerodynamics and physical principles. How it does this is not immediately obvious. Modern science eventually learned the secret of the boomerang. This knowledge only adds to the mystery of how humans invented the boomerang in the first place.

That mystery grew even deeper several years ago, when a boomerang carved out of mammoth tusk was discovered in Poland. The 27-inch long boomerang had a carefully carved airfoil every bit as sophisticated as any Australian boomerang. While this is the oldest boomerang ever found, boomerangs have been discovered on four continents. The American Indians of the Southwest once used boomerangs, as did those in Alaska. And a gold-tipped boomerang was discovered in the tomb of King Tut.

These discoveries offer a powerful witness that humans are unique. It appears that they once understood the sophisticated aerodynamic principles used by the boomerang. The universal use of the boomerang strongly suggests a common history for all humanity that is more recent than many scientists care to admit. The genius of the boomerang tells us that humans have always been distinctly human—highly intelligent and creative.

Prayer: Dear Lord, I thank You for the witness of artifacts that show that humans have always been special creatures. Especially protect our Christian young peoples' faith from the lie that humans came from ape-like creatures. Help them seek their identity in You and the forgiveness of sins You won for us. Amen.

REF.: Mammoth boomerang. *Discover*, Mar. 1988. p. 18.

Did Plants Study Insects?

Isaiah 40:8

"The grass withereth, the flower fadeth: but the word of our God shall stand for ever."

People universally appreciate flowers. Their myriad colors and sweet fragrances fill flower gardens and vases around the world. All of this makes evolution's explanation of why there are flowers even more interesting.

According to evolution, flowers developed long before there were bees, moths or butterflies to pollinate them. They tried to solve this problem by suggesting that perhaps beetles pollinated flowers. This suggestion makes another chapter in many books on evolution all the more interesting. I'm referring here to the chapter on how flowers are designed to keep non-pollinators, like beetles, away from the nectar.

We are also told that the beauty we find in flowers is quite accidental. To quote one book, "Indeed, if insects did not have color vision, flowers would be drab and if insects could not detect odors, flowers would have no fragrances." This leads us to ask how flowers knew that insects could see colors and smell sweet odors. Even if plants could study insects and share their findings with each other, they still could not engineer changes in their design. Humans have wanted to fly for thousands of years. But for all the wishing, no one has yet grown a feather or a propeller!

Flowers and the complex interrelationships through which they are pollinated form one of the most beautiful and marvelous examples of the Creator's work. Perhaps that's why it is so difficult for evolution to explain them!

Prayer: I thank You, Lord, for the beauty that surrounds us every day. I marvel that even the destruction that sin has brought into the world has been unable to destroy the beauty of Your work. Help me to remember Your greatest work - my salvation through the forgiveness of sins, for this destroys sin, now and forever. Amen.

REF.: Camazine, Scott. 1987. *The Naturalist's Year*. New York: John Wiley & Sons, Inc. p. 86.

Which Is Smarter, Bird or Ape?

Genesis 1:26

"And God said, Let us make man in our image, after our likeness: and let them have dominion over the fish of the sea, and over the fowl of the air, and over the cattle, and over all the earth, and over every creeping thing that creepeth upon the earth."

In 1917 a German scientist decided to test chimpanzees for abilities beyond intelligence. He wanted to find out whether chimpanzees could take unrelated facts and put them together to solve a problem—an ability we call insight.

For the test, the scientist suspended a banana in a cage out of reach above the chimps. He then placed a box on the floor of the cage 10 feet from the banana. What happened next astounded the scientist. The chimps quickly learned that they could not get to the banana, even with their highest leaps. So they went to the box, pushed it under the banana and climbed on the box to take the banana. Their conclusion was that these chimps are so clever because they are closely related to us on the evolutionary tree.

Recently a Harvard scientist tried the same experiment using pigeons and a bird-size, fake banana. The pigeons had earlier been taught a number of tricks, including banana pecking and box climbing. These tricks had been taught to them as unrelated activities. However, when faced with the same problem as the chimps, they moved the box under the banana, climbed on the box, and pecked the banana. Pigeons, too, have insight!

One can only conclude from this study that either insight among chimps has nothing to do with evolutionary relationships, or that maybe we have evolved from pigeons! Seriously, the real lesson here is that similarities between creatures have nothing at all to do with any supposed evolution.

Prayer: Dear Lord, I thank You for Your Holy, innocent suffering and death, and Your resurrection, so that I might be raised up as a new creature, recreated in Your image. Amen.

REF.: Bird brain. *Discover*, May 1984. p. 12.

Love with Eight Arms

Matthew 23:37

"O Jerusalem, Jerusalem, thou that killest the prophets, and stonest them which are sent unto thee, how often would I have gathered thy children together, even as a hen gathereth her chickens under her wings, but ye would not!"

While many people think the octopus is an ugly creature, the octopus is intelligent. The mother octopus puts a great deal of love and effort into the care of her young.

Fertilized octopus eggs are about the size of a grain of rice. They are connected in clusters on slender stalks. Once her eggs are ejected and fertilized, the mother octopus hangs the threads containing her eggs from the roof of her home. It may be a rock cave or even a dark, enclosed man-made structure. By the time she is finished hanging her eggs, she will have almost 20,000 ivory capsules strung around her. Now her real work begins. Until the eggs hatch, mother octopus will give up food, concentrating on nothing but cleaning her eggs of any material that might result in a fungus or parasite infection. She cradles the eggs gently in her arms, brushing each of the 20,000 eggs, and sometimes washing them with jets of water. There is nothing else she can do for her young, but this care receives her single-minded attention.

It has rightfully been said that one cannot create something that one doesn't have. If one does not have love, one cannot create love in others. Therefore, the Creator of the octopus must know a great deal about giving individual love and care to a multitude of individuals. The Bible is our Creator's own personal description of His love for you and me. Whether you have experienced that personal care and love of your Creator or not, you can certainly learn the depth of His care for you through Jesus Christ in the pages of the Bible.

Prayer: Dear Father, not only are Your skill and wisdom evident in the creation, but we also see hints of Your love as well. Help me to realize more completely what Your personal and intimate love for me really means in my day-to-day life. In Jesus' Name. Amen.

Spring-Operated Birds

Genesis 1:20

"And God said, Let the waters bring forth abundantly the moving creature that hath life, and fowl that may fly above the earth in the open firmament of heaven."

Did you ever wonder just what that wishbone in a turkey, chicken or duck was for? You may have gotten a hint about its purpose if you ever tried to snap the bone before it was dry.

Amazing as it sounds, until recently, even scientists didn't know what the wishbone was for. They're still not sure exactly what all it does.

Scientists placed starlings in a wind tunnel. The wind tunnel was designed to allow the birds to fly without moving too much in relation to the tunnel walls. The scientists took x-ray movies of starlings flying in place. As a result, they learned that the wishbone is a spring. When the starlings are at rest, the spread between the open ends of their wishbones is only 11 or 12 mm—about half an inch. However, when they open their wings and fly, the wishbone opens up to 18 or 20 mm—about 3/4 of an inch. The wishbone opened and closed in this way with each wing stroke.

Scientists aren't sure whether the wishbone helps in flight. They suspect that the wishbone greatly expands the lungs of the bird so that it can take in greater amounts of air. In this way, the bird doesn't become winded so easily as it flies. To use modern engineering language, the wishbone spring inside the bird acts like a second, and much higher capacity, breathing system. Its effect is to turbo-charge the bird. This obvious design makes it even more reasonable to accept the idea that the animal world is the work of an ultimate engineer—the Creator who sent His Son to redeem us.

Prayer: I joyfully thank You, Lord, that there is no end to Your creativity nor limit to Your wisdom in designing the creation. Amen.

Ref.: *Star Tribune,* September 16, 1988 –AP

A Community in the Air

Psalm 148:7a, 9-10

"Praise the LORD from the earth, . . . Mountains, and all hills; fruitful trees, and all cedars; Beasts, and all cattle; creeping things, and flying fowl:"

Much of the world's surface is covered by rain forest. If you are unfamiliar with the rain forest, you may be tempted to compare it to forests you have seen. But really there is no comparison at all.

In a rain forest, the earth's living surface is covered by a second living habitat some 50 to 150 feet in the air. Here, thousands of species of plants and animals live and breed for generations without ever touching the ground.

The canopy of the rain forest is supported by huge trees that usually have no branches for the first 50 or more feet. They grow so close together and their tops are so dense that sunlight seldom reaches the jungle floor. This canopy can be up to 100 feet thick. While you might expect to find monkeys living in the canopy, you may not expect to find mice, frogs, and even earthworms there as well. Here, most of these creatures are born, they live and die without ever touching the ground. The rain forest canopy is truly a second environment for life high above the surface of the earth!

Unusual environments like the rain forest canopy show us that the creativity of God is unlimited. He can create living things to carry on successful lives in any environment. He can create varied communities that interact and depend upon each other within each unique environment. His creation is truly a wonder!

Prayer: Dear heavenly Father, all of creation, in familiar forms, amazing forms and life I cannot imagine praises You! I praise and thank You, too. In Jesus' Name. Amen.

Ref.: *Science digest,* October, 1978, page 30

Bats Who Feed Trees

Psalm 104:27-28

"These wait all upon thee; that thou mayest give them their meat in due season. That thou givest them they gather: thou openest thine hand, they are filled with good."

The surface of the earth 50 to 150 feet below the great living canopy of the rain forest is a dark, humid, still world dominated by great columns of tree trunks. Within those trunks and some of the giant hollow branches extending from them lies the secret of the life of the canopy itself.

A rain forest typically receives 12 or more feet of rain per year. That much rain washes out of the soil most of the nitrates needed by the trees for growth. What the trees need is a rich source of nitrates that is constantly being replenished. And that's where the bats come in.

Typically, these huge trees are hollow inside. Many different kinds of creatures, including fruit bats, enter the hollow trees through various openings. Fruit bats find ideal daytime sleeping rooms inside the great hollow branches that extend like caverns from the hollow tree trunk. The accumulating layer of bat guano inside the tree itself is one of the richest sources of nitrates known. So the bats provide the tree with the nitrates it needs in exchange for a protected home during the day!

In a very real sense, the fruit bats are the collection and transport system for the raw materials that make possible the tropical rain forest canopy with its millions of residents. The Creator has devised an ingenious way to provide for the needs of many creatures. Truly the Lord does provide all living things with their food in precisely the way they need it!

Prayer: Dear Lord, I pray that You would fill me with as much love as You have for the creatures of Your wonderful creation. And just as the entire rain forest must look to the fruit bat for life, help me always to look to my Savior, Jesus Christ, for all life. Amen.

Ref.: *Science digest,* October, 1978, page 30

The Very Hairs of Your Head

Matthew 10:30

"But the very hairs of your head are all numbered."

In showing how the Creator is intimately involved with His creation, Jesus Christ said that the very hairs of our heads are numbered by Him. This means that no detail is too small to escape His attention; no change misses His careful and loving eye.

The hair that you see is nothing more than dead protein that is produced by the cells of the hair follicle anchored inside the layers of the skin. The total number of hair follicles on the adult body is about 5 million; only about 100,000 of them are on the scalp. Each hair grows out of one follicle for about three to five years. Then the hair is shed and the follicle rests for about three months before starting to grow hair again.

So you see, once you know how many hairs are on your head, it's still no easy job to keep track of their changing number. The average scalp hair grows about one inch every two or three months. This means that each day your head is growing the equivalent of one hair, 100 feet long—that's nearly seven miles per year!

Yes, it's true, the Creator cares so much about you that He knows from moment to moment how many hairs are on your head. He has not created our world and then left us to drift through space and life all alone. He has even expressed His love and purpose for you in the words of the Bible. Take a look today at what Scripture says about His Son, who loved you so much He even gave His life for you!

Prayer: I confess, dear Lord, that I cannot understand how You can keep track of all the details of the creation as Your Word says You do. But You are God and I am only human. Don't let my weakness limit or weaken my faith in Your day-to-day involvement in my life. In Jesus' Name. Amen.

Ref.: *Science Digest,* June, 1979, Page71

Can Evolution Predict the Future?

Psalm 39:11

"When thou with rebukes dost correct man for iniquity, thou makest his beauty to consume away like a moth: surely every man is vanity. Selah."

We have often pointed out the many plants and animals that are interrelated. Their lives depend upon one another. Very often they are interrelated in such a way that creation offers the only reasonable explanation for their special features. Evolution can call on "lucky coincidence" just so many times to explain these relationships. It makes far more sense to say that all living things were carefully planned and designed by an all-wise Master Designer.

One interesting example is found in the moth family *Noctuidae*. This family of moths has special organs that detect the ultrasonic sounds made by the bats who eat them. A "lucky coincidence"? The problem with the evolutionary explanation is this moth is dated by evolutionists to 75 million years ago. (Remember, these are evolutionists' years, not real years.) This is back during the time of dinosaurs, when there weren't supposed to be any bats. According to evolution, bats didn't arrive on the scene until 55 million years *after* the *Noctuidae* moth. One wonders whether mindless evolution could see into the future. Or perhaps evolution was able to make moths with an ability to hear bats that would not exist for the supposed tens of millions of years! Neither explanation sounds very scientific!

It makes a good deal more sense to throw out the millions of years of inflated evolutionary time. We can view the unique design of the *Noctuidae* moth as a sign from an all-knowing Creator that all living things were made by Him. Only an intelligent, personal Creator can account for the complex, overall plan that we see in the created relationships between vastly different creatures.

Prayer: Father, the scope and complexity of the interrelationships in nature shows how poor our most complex projects truly are. Help me to keep my life in perspective so that my dependency is always on You. In Jesus' Name. Amen.

Frog Hibernation

Exodus 8:10b-11

"And he said, Be it according to thy word: that thou mayest know that there is none like unto the LORD our God. And the frogs shall depart from thee, and from thy houses, and from thy servants, and from thy people; they shall remain in the river only."

Did you ever wonder where frogs go in the winter? Yes, frogs hibernate, but there is much more to it than that.

Frogs are cold-blooded, meaning that they don't create their own body heat. The heat they need to maintain life must come from the sun or the heat in the water and air around them. As the weather cools off, a frog's blood begins to thicken and life slows down. Frogs begin looking for places to hibernate. The frog must carefully select his winter resting spot. If his body temperature falls more than two degrees below the freezing point of water, ice crystals will begin to form in his blood, and he will die. Some frogs bury themselves deeply enough in the mud to avoid freezing temperatures. Others may crawl inside decaying logs or under leaves, both of which provide heat from decay to avoid freezing temperatures.

Scientists have found that frogs have still another trick to help them survive. When the weather is unseasonably cool during the spring or summer, meaning a harder life and more deaths among the frog population, the next generation of frogs will automatically have more females. Additional females means that more eggs can be laid in the spring. As a result, the frog population will more quickly return to normal.

Everywhere we look in the creation we find evidence, not only of the Creator's wisdom, but of His loving care for every living thing. Do you know His loving care in your life? You can be absolutely certain of that care through the forgiveness of sins in Jesus Christ.

Prayer: Dear heavenly Father, all creation serves and honors You. Help me to conduct my life as Your redeemed child so that I, too, may better serve and honor You. In Jesus' Name. Amen.

Ref.: *Science Digest,* November, 1978, Page 96

The Original Antibiotic

Genesis 1:28

"And God blessed them, and God said unto them, Be fruitful, and multiply, and replenish the earth, and subdue it: and have dominion over the fish of the sea, and over the fowl of the air, and over every living thing that moveth upon the earth."

In Genesis 1:28 God tells humans to subdue the earth. That command implies that He has placed within creation tools and knowledge that humans can learn to use for their own benefit.

After sin entered the world and sickness and death became a reality, God's remarkable foreknowledge became evident. One of the first antibiotics ever discovered by humans has been in use for thousands of years. Modern researchers are just beginning to appreciate the wonder of this natural antibiotic that kills some 650 different strains of disease organisms. It is virtually nontoxic. And best of all, disease organisms don't become resistant to it.

What is this miracle antibiotic? Silver. Ancient Greeks and Romans used silver containers to keep liquids fresh. American settlers would often place a silver dollar in milk to delay its souring. Most of the world's airlines use silver filters on aircraft to prevent dysentery. After testing 23 different methods for purifying water, NASA selected silver water filters for use on board the space shuttle. Japanese researchers have found that silver is even able to detoxify some poisons.

God's knowledge is far above ours. He has hidden in the creation around us knowledge that can help us in this earthly life. He invites us to learn from Him. He has "hidden" greater knowledge, knowledge that can help us here and in eternity, in the Bible. How could anyone be satisfied with what little we know from His great mind!

Prayer: I am filled with awe, dear Lord, at Your great mind. But I am even more overcome by Your love for me in Your innocent suffering and death on the cross for my sins. Thank You! Help my appreciation for this gift to grow forever! Amen.

Ref.: *Science Digest,* March, 1978, p. 57-59

Parents Who Are Wise as Owls

Ephesians 6:4

"And, ye fathers, provoke not your children to wrath: but bring them up in the nurture and admonition of the Lord."

The screech owl is one of the smaller and more inoffensive members of the owl family. While most owls have a large variety of calls and screeches, the screech owl is one that seldom screeches.

Usually the screech owl is little bother to anyone except its lunch. However, when it is time to raise a family, the personality of the screech owl changes and they can upset many lives. Screech owls don't build nests, they simply look for a good sheltered place to lay their eggs and protect their young. This could be a natural hollow in a tree or an abandoned woodpecker nest. After about 28 days, the eggs hatch into hollering baby screech owls.

The parents know that their work is cut out for them now. If they don't provide enough food for their youngsters, the hatchlings will eat each other right in the nest. So parents pay single-minded attention to their young's needs. They will raid area bird houses and even attack people. In one instance, a screech owl came down a chimney and yanked a canary out of its cage to take back for its hungry youngsters.

It seems that the normally inoffensive screech owl will use all of its resources to provide for its young. Has the Lord provided us with a reminder in the screech owl about how we should place our own children on a little higher priority in our own busy lives?

Prayer: Dear Father, I ask that You would help our earthly fathers be more like You. Instill in Christian parents a deeper sense of the need to complete their task of bringing up their children in the training and admonition of the Lord. In Jesus' Name. Amen.

Ref.: *Science Digest,* March, 1979, Page 62

Plant Self-Defense

Job 38:4

"Where wast thou when I laid the foundations of the earth? declare, if thou hast understanding."

Every day, all over the earth, plants engage in chemical warfare against insects and animals that would eat them. When this drama is described by evolutionary scientists, they usually talk about plants as though the plants were skilled chemists who developed their abilities on their own.

For example, the leaves of the oak tree contain tannins. Tannins form complexes with proteins so that, when eaten, they have little nutritional value. An evolutionary account of how this arrangement came about describes how the trees supposedly developed this strategy for self-defense. Stories like this make one wonder how oak trees gained so much knowledge about animal digestion and chemistry. Some species of milkweed and dogbane produce powerful muscle relaxants that can be fatal to humans.

One might picture, in the far distant past, white-smocked milkweeds working in the chemistry lab. Other "doctor" milkweeds are feeding various concoctions to humans in cages to test their responses. When one tries to account for this without a Creator, the picture can become silly.

The Bible provides an answer that makes a great deal more sense. In Job 38 and 39, the Lord—the Creator—asks Job about dozens of aspects of the creation. In His questioning, He asks Job about the source of the knowledge and abilities found in living things. The answer, of course, is that the creating God, and no one else, designed, built and taught the creation. Other answers simply can't satisfy.

Prayer: Father, remind me that You are the source of all true knowledge and wisdom in the world, especially when I am tempted to follow human wisdom. In Jesus' Name. Amen.

Ref.: *Science Digest*, March, 1979, Page 66

Plants Skilled in Animal Chemistry

Genesis 1:11

"And God said, Let the earth bring forth grass, the herb yielding seed, and the fruit tree yielding fruit after his kind, whose seed is in itself, upon the earth: and it was so."

It occurred to scientists only about 30 years ago that insect pests might be controlled through hormones. If, for example, a hormone that prevented a destructive insect from reaching maturity was applied to a crop, the interrupted life-cycle of the insect might prevent damage to the crop. Yet the hormone would be a safe, natural substance.

However, scientists found that synthesizing such hormones was much more difficult than they thought. Several years into the research, Japanese scientists discovered that certain plants were already making these hormones for self-defense. In fact, many of the plants were manufacturing what they called "super-hormones." Super-hormones are hormones that work in far lower concentrations than the juvenile hormones made by the insects themselves. These juvenile hormones prevent insects from reaching their adult, and usually most destructive, stage. Among those plants that make juvenile insect hormones are ferns and certain evergreens.

One result of this research is the development of a mosquito juvenile hormone. This hormone can be time-released from briquettes in swamps, preventing mosquito wigglers from ever maturing to their flying, biting, adult stage.

Again, the incredible design and balance that could only have been built into the world by a wise Creator are evident in the creation. How else could plants have gained such skill in animal chemistry?

Prayer: Dear Lord, guide our thinking so that we may discover the wisdom You built into the created world. Help us to learn how You have made things so that we may find nondestructive ways to control pests. Ultimately, help me and all Christians to tell those around us about the greatest wisdom of God, which is found in the Gospel of salvation. Amen.

Ref.: *Science Digest,* March, 1979, Page 67

Nature's Velcro

Ecclesiastes 1:9

"The thing that hath been, it is that which shall be; and that which is done is that which shall be done: and there is no new thing under the sun."

The loop-and-hook fastener called Velcro has come a long way since it first appeared. Velcro goes into space as a light, easy-to-use fastener. Here on earth it's replacing everything from laces in children's shoes to buttons and zippers on coats.

However, humans were not the inventors of the hook-and-loop fastener. Plants were the first to use this principle. Scientists learned about the principle when they were checking out stories of how bean leaves were used to trap bedbugs during World War II. They discovered that the hooked hairs on the underside of bean leaves did indeed trap bedbugs by their feet. The bedbugs stayed stuck to the leaves until they starved to death.

In nature, the hooked hairs trap leafhoppers, leaving them unable to munch on the leaves of bean plants. Further research has shown that some varieties of wheat trap the cereal leaf beetle, preventing damage to their leaves.

Most amazing of all is the timing of this defense. Plants that use this method to protect themselves don't start growing the hairs until a few days before the annual attack by their insect enemies!

Humans might think their inventions are clever. However, because God is smarter than we are, it is very likely that somewhere, somehow, He has already created nearly every practical invention. This is yet another argument for creation!

Prayer: I thank You, Father, that in commanding us to subdue creation, You guided us to discover what You have already invented. Help our technology to glorify You by giving Christians who are working in the sciences insights into Your inventions. In Jesus' Name. Amen.

Ref.: *Science Digest,* March, 1979, Page 68

Chemical Communication

Psalm 40:5

"Many, O LORD my God, are thy wonderful works which thou hast done, and thy thoughts which are to us-ward: they cannot be reckoned up in order unto thee: if I would declare and speak of them, they are more than can be numbered."

There are many methods of communication, some of which we human beings don't often think about. One of these is chemical communication.

Take, for example, the female silkworm moth. When she is ready to mate, she releases a very small amount of a chemical. This chemical, called a *pheromone*, can attract a male silkworm moth from up to two miles away. Even though the female has released only 100-millionth of a gram, the male can sense the chemical. Many other insects use similar pheromones to attract mates.

Pheromones are also used for chemical communication by ants to mark their trails. When an ant finds a good source of food, it returns to the nest, leaving an invisible scent trail. Its fellow ants also use this trail to find the food. As long as they still find food, they will continue to reinforce the scent. Once the supply is exhausted and ants no longer follow the trail, it disappears within a few minutes. This chemical scent is so specific that another species of ant will not recognize the trail if he stumbles upon it. This helps protect the ants' food source from being discovered by other ants.

Chemical communication allows the communicator to communicate to another of its species, even over long distances. In one sense, pheromones are the animals' version of radio.

Prayer: If it had been up to me to design and build the creation, dear Lord, my best effort would not compare with what we have from You. Help me to remember this when I try to build my life without You. Amen.

Ref.: Emmel, Thomas C. Worlds Within Worlds-An Introduction to Biology, p. 333

Tiny Time Bombs

Colossians 1:16

"For by him were all things created, that are in heaven, and that are in earth, visible and invisible, whether they be thrones, or dominions, or principalities, or powers: all things were created by him, and for him:"

The discovery of tiny pollen grains in rock shows just how inflated the billions of years claimed by evolutionists really are. This discovery, by a creation scientist, has turned those pollen grains into tiny time bombs that continue to damage the theory of evolution.

In 1964, Dr. Clifford Burdick devised a new, better way to collect pollen samples, and he used this method to gather samples in the Grand Canyon. These pollen samples, when viewed under the microscope, allow scientists to identify what kinds of trees they came from. The Hakatai shale, dated by evolutionists at 600 million years old, was among the layers Dr. Burdick examined for pollen. According to evolutionary dating, this layer would not have contained pollen from conifer trees. Conifers supposedly hadn't yet evolved. When conifer pollen was discovered in the samples, evolutionists quickly claimed that they were modern pollen grains that had contaminated the sample.

The problem with the evolutionists' explanation is that pollen is resistant to staining. Yet this pollen was stained with red from the shale itself. It must have been there a long time. The experiment was repeated in 1970. Not only were grains of conifer pollen again found, but pollen from oak trees and evidence of other flowering plants were found as well.

The discovery of these pollen grains offers scientific evidence that evolution's time scale and story of history is wrong. The evidence does support the accuracy of the Bible that tells us that all life began during the first week of the earth's history some few thousand years ago.

Prayer: I thank You, Lord, for those scientists who work to honor You. Bless them and keep them under your protection as they work in a field that is generally hostile to your Truth. Amen.

REF: Burdick, Clifford L. 1981. Pine pollen in the Grand Canyon. *Bible-Science Newsletter*, June. p. 1.

How to Talk Gerbil

John 1:14

"And the Word was made flesh, and dwelt among us, (and we beheld his glory, the glory as of the only begotten of the Father,) full of grace and truth."

Gerbils are those friendly, furry little rodents that many children keep as pets. Gerbils are social animals. They value communication very highly. Researchers have found that those gerbils in a given community that can communicate most clearly have the highest social status.

For a long time, researchers were confused about the sounds gerbils make, because they were only hearing a small part of the communication. This is because our human ears hear sounds that go up to about 20,000 cycles per second. However, gerbils do most of their talking at as high as 32,000 cycles per second. Their talk is too high-pitched for us to hear.

Tape recordings of these ultra-high-frequency sounds have been played back at slower speeds. They reveal that most gerbil communication involves different kinds of whistles. These whistles are different types of speech, depending upon the purpose of the communication. While researchers cannot completely translate gerbil speech, they have learned that gerbils are continually asking where others are and reporting their own positions.

Scripture calls the Son of God the Word. He is presented in the Bible as the creating and saving Word. This revelation about Him is intended to help us realize that communication is an essential part of God's nature. Communication is not an evolutionary development. How could it be? Communication is the gift of a loving Creator. Have you read His communication to you in the Bible?

Prayer: Father, I confess that my life is filled with too many words, yet not enough of the Word. Forgive me for Jesus' sake, and help me keep my resolve to spend more time being instructed in Your Word of truth and life. Amen.

Ref.: *Science Digest* March 1978 Page 53

The Gift of Humor

Ecclesiastes 3:1, 4

"To every thing there is a season, and a time to every purpose under the heaven. . . A time to weep, and a time to laugh; a time to mourn, and a time to dance;"

When Scripture talks about laughter, it most often speaks of laughter as a form of derision. God is described as laughing at the pride and plans of the wicked. However, that does not mean that God is against *the right kind* of humor.

After all, He gave us the ability to laugh. In Ecclesiastes 3:4 we read that there is also a time to laugh. Jesus reminds the faithful, in the Beatitudes, that those who weep now will laugh with joy later. When you stop and think about it, anyone who can create the playful kitten cannot be against laughter.

Medical researchers are finding out just what a wonderful gift laughter is. Laughter, it has been found, is excellent exercise for the lungs. It clears the respiratory system and is emotionally healthy. Laughter relaxes muscles. It is a proven reliever of emotional tension. Research also suggests that laughter also has healing effects, possibly boosting the immune system.

Laughter also has been found to have a natural pain-killing effect. In one experiment involving patients viewing funny films, the pain-killing effect of laughter lasted for up to two hours.

Laughter has also been identified among chimpanzees and apes. Very possibly, other creatures have their own way of laughing. If we remember that Scripture also encourages us to "avoid foolish talk," we can also give thanks to God for the gift of humor.

Prayer: I thank You, Father, that You are not a grim God who wants us to be somber and joyless. I also thank You for the humorous things You have made. Grant me good, healthy joy and humor in my life. In Jesus' Name. Amen.

REF: Peterson, K. 1996. A chuckle a day does indeed help keep ills at bay. *USA Today*, Oct. 31. p. 10D.

The Gift of Smell

Ephesians 5:2

"And walk in love, as Christ also hath loved us, and hath given himself for us an offering and a sacrifice to God for a sweetsmelling savour."

The ability to smell is one gift we often take for granted. That's probably because we usually identify things more quickly with one of our other senses. At the same time, Thanksgiving dinner wouldn't be the same without the smells.

Some of the most unlikely creatures have a sense of smell. Believe it or not, even fungi have a sense of smell. Worms have organs on their heads to sense odor. Ticks carry their scent-detecting organs on their feet; this arrangement wouldn't work well for us! Mollusks smell through their gills! The salmon uses smell to find the same brook in which he was spawned. Lizards and snakes use their tongues to detect scents.

The most skilled smellers in the animal world can detect a scent even if there is only one molecule of that scent mixed in with 10 trillion molecules of average air. The modern perfumer's highly trained nose can identify about 10,000 different odors.

Before the days of modern medicine, a doctor's diagnosis would consist in part of sniffing the patient. For example, the plague had the scent of honey. Scarlet fever often gave its presence away by the smell of hot bread, and measles smelled like fresh-plucked feathers.

It takes a keen knowledge of biochemistry to design a system that can detect and identify scents. So next time you smell a wonderful meal or a beautiful flower, you may want to thank your Creator.

Prayer: I thank You, Lord, for my senses and all the wonderful scents with which You have filled the world. Help my life to be a sweet-smelling offering ascending to You. Amen.

Ref.: *Science Digest* June 1979 Page 45

God's Unlimited Generosity

Psalm 35:18

"I will give thee thanks in the great congregation: I will praise thee among much people."

There is nothing our Creator God does that is not intended to bring us closer to Him. Why? Because His purpose in making humans was to make someone to love. We exist because He wanted to love us and be loved by us.

There is nothing visible or invisible that is not meant to be of service to us. Consider the invisible angels that protect us. They rejoice when even one person turns to faith in Jesus Christ as Lord and Savior.

Consider the heavens above our heads. There, God placed billions of stars—some much larger than our sun. There are giant pinwheel galaxies with millions of huge stars. Countless billions of stars churn out more energy every second than we can ever hope to measure. What a lavish and beautiful timepiece to tell us when it is time to plant and time to harvest! God has given us more than the necessities of life. He has also hidden medicines within the bark of trees and the sweet-scented flower. He has filled that creation with splendor and beauty.

However, I cannot neglect to mention His greatest and most generous gift. When we sought our will instead of His, we withdrew our love from Him. His response was to send His only Son, Jesus Christ, to suffer our punishment for sin on the cross. And He rose from the dead on the third day so that we could have new life with our Creator again. Oh, that our thanksgiving to God could be as unlimited as His generosity!

Prayer: How can I offer, dear God, a gift that is suitable to show my thanksgiving to You for Your generosity to me? Even my life is a pittance compared to Your gifts. Yet that is all You ask. Fill me with joy in giving You my life. In Jesus' Name. Amen.

The Efficient Firefly

Psalm 50:2

"Out of Zion, the perfection of beauty, God hath shined."

There are more than 200 species of fireflies around the world. Almost everyone has seen these small insects flying about or sitting in the grass. On dark summer nights they can be seen flashing their lights like ships far out at sea. As they flash, they are communicating very specific messages to one another.

The light created by fireflies is called cold light because it is produced without creating any heat. To make a flash, the firefly must make and mix a chemical called *luciferin* with oxygen and an enzyme called *luciferase*. This mixture is combined with a catalyst to create the flashes of light. The result is a method of producing light that is far more efficient than any source of light in your home. A relatively tiny amount of energy creates a generous amount of light.

Fireflies do not make light so they can read a book. Their flashes are used for finding a mate. Fireflies are able to read the light signals of other fireflies. These signals may contain several messages. Some flashes can tell a firefly whether the sender is one of his species. Light flashes are used to identify whether the sender is a male or female. If it's a female, the flashes can identify whether she has already mated.

This is another example of the unlimited creativity and extravagance of our Creator God. The same beauty of the firefly that creates wonder in the child should also fill us with wonder at the limitless imagination of God.

Prayer: Lord Jesus Christ, You are the Light of the world, and Your saving work for me is the light in my life. Help me better appreciate the beauty You have made. Then help my appreciation of that beauty be given words that are an effective witness to others. Amen.

Ref.: *Science Digest,* August, 1978, Page 59

How to Speak Firefly

John 8:12

"Then spake Jesus again unto them, saying, I am the light of the world: he that followeth me shall not walk in darkness, but shall have the light of life."

Those light flashes created by fireflies are actually a quite complex language. The more than 200 species of firefly each have their own language. Each species uses a different combination of color, intensity and flash intervals.

As the male firefly cruises over the firefly community, he emits an automatic series of flashes. As he "talks" with flashes of light, he is looking for the appropriate response from a female who is ready to mate. Each species has its own light language, helping males to identify members of their own species. Females who are not ready to mate or who have already mated communicate these facts to the overhead male. Evidence indicates that at least one firefly species understands and can speak "foreign" firefly language. Females of this species will lure males of other species down by using the male's language and pretending to be a female who is ready to mate. When the male comes close for mating, the female captures and eats him.

You can "speak" firefly, too, if you have a flashlight. First you must cover the flashlight so that it emits only a tiny point of light. Then, after carefully watching the patterns of the female fireflies in the area, try to mimic that pattern. If you do this correctly, male fireflies will come to investigate your flashlight.

The intelligence that fills all of creation is one of the "fingerprints" of our intelligent Creator. Such intelligence cannot come from accidents.

Prayer: Lord Jesus, You are the light of the world. Instill within me a deeper appreciation for the gift of forgiveness of my sin so that I will find it easier to shun the darkness of the world and walk in the light of Your grace. Amen.

Ref.: *Science Digest,* August, 1978, Page 59

Could Adam Have Lived that Long?

Genesis 5:5

"And all the days that Adam lived were nine hundred and thirty years: and he died."

In Genesis chapter 5 we read that Adam lived to be 930 years old. One question often asked is, "Could Adam really have lived that long?"

Most of the effects of aging are due to genetic mistakes that accumulate over generations. This is in addition to the age limitations God imposed on human beings after the Genesis Flood. These early generations, so close to Adam and Eve, would have accumulated comparatively few genetic mistakes.

Before the Genesis Flood, the earth was probably a very different place. Many creation-scientists believe that the pre-Flood earth was surrounded by a canopy of water vapor or at least in some way had a greater atmospheric pressure—perhaps about twice what it is now. Medical researchers have been experimenting with the effects of higher atmospheric pressure. Tests in chambers that subject people to similar air pressures show that high atmospheric pressure gets more oxygen to the cells and speeds healing. Hospitals use a hyperbaric chamber for burn victims. Experiments have shown that such chambers can also reverse senility. And higher atmospheric pressure appears to reverse some of the effects of aging on the skin and hair and stimulates the entire body.

We can be sure that Adam and the others lived as long as they did because God's Word says they did. However, science may be discovering some of the conditions in the pre-Flood world that may have helped make these long lives possible.

Prayer: Father, I trust Your Word because I know that You could not lie or mislead, even if You wanted to. Even as I confess my trust in the truth of Your Word, let me not neglect studying it. Grant me Your Holy Spirit so that my understanding of it would grow. In Jesus' Name. Amen.

Ref.: *Science Digest,* August, 1978, Page 33

The Greatest Deep-Sea Divers

Genesis 1:21

"And God created great whales, and every living creature that moveth, which the waters brought forth abundantly, after their kind, and every winged fowl after his kind: and God saw that it was good."

Even though it's a mammal and not a fish, the Wedell seal of the Antarctic is one of the greatest deep-sea divers in the world. Weighing up to 1,200 pounds, the Wedell seal can cruise under water for up to 70 minutes. It is able to reach depths as great as 1,600 feet. For comparison, the record for a human being is a 13-minute, 43-second dive down to 282 feet.

Humans are not designed to handle the pressure changes involved in diving to great depths. If humans stay too deep for too long and then return to the surface too quickly, they may die. The rapid change in pressure causes nitrogen to come out of solution in the blood. The result is a potentially fatal condition known as the "bends."

Wedell seals don't have to worry about this, because their blood doesn't have dissolved nitrogen. Scientists are trying to learn just how the Wedell seals' lungs prevent this. However, the unique design of the Wedell seal doesn't stop with its lungs. The Wedell seal is able to withstand pressures up to 750 pounds per square inch. Those pressures on you or me would cause us to have epileptic seizures.

What science is learning about the complex and unique designs of the Wedell seals makes the evolutionary claims that these seals came from land-living mammals increasingly difficult to believe. Yet they illustrate the great wisdom of our Creator!

Prayer: Dear Lord, You have created the mountains and the depths of the seas. Where You have commanded creatures to live, there they live to Your glory. Help me to live the life You have given me to Your glory. Amen.

REF.: Browne, Malcolm W. 1984. Antarctica: life's tenuous toehold. *Discover*, Mar. p. 44.

The First Five Years of Creation

Psalm 44:1

"We have heard with our ears, O God, our fathers have told us, what work thou didst in their days, in the times of old."

How old did the earth look five years after the creation? Geologists tell us that it takes hundreds of thousands of years to wear down new rock into sand and soil and for the sea to cut a beach. If this was true, one would not expect to see beaches or soil on an earth that was only five years old. Or would we?

Off the coast of Iceland, over 300 feet below the surface of the ocean, hot volcanic rock began to seep into the cold ocean water in 1963. By 1967 over half a cubic mile of new, hot rock had formed a brand new island where there had once been only ocean. This new island was named Surtsey.

The following summer, scientists arrived on the island to study it. They found that a basalt cliff on the beach had been eroded over 300 feet inland by the sea in less than five years. They found that several parts of the island already had terraced beaches up to 300 feet wide. On the north side of the island, they found that newly formed sand had formed a sand bar that extended almost 1,000 feet out from the original island.

If geologists had not known that Surtsey was less than five years old, they would have guessed that the island was thousands or even hundreds of thousands of years old. After all, many geologists, rejecting the biblical account of a young creation, have used features such as sand beaches to try to prove that the earth is millions of years old!

Prayer: I thank You, Lord, for the record of Your deeds that has been preserved for our learning in the Bible. I thank You for the sacrifices of those who preserved it and for those who gave their lives for its truth. Help me to live a life worthy of the great gift of Your saving Word. Amen.

REF.: Normann, J.O. 1970. Trends in post-volcanic development of Surtsey Island. *Surtsey Research Progress Report V* (Reykjavik: Surtsey Research Society). pp. 95-112.

One of the Strangest Plants

Job 26:12

"He divideth the sea with his power, and by his understanding he smiteth through the proud."

One of the largest food crops in the world grows in the icy blue light that penetrates eight feet of Antarctic ice. These plants, called diatoms, grow attached to ice crystals in sub-freezing temperatures. Their cell walls are not made of cellulose, as are the cell walls of other plants. Rather, they are made of silica, the same material that makes up much of the world's sand.

Like other plants, diatoms photosynthesize. However, they carry out their work in light that is less than 1% as bright as light at the surface of the Arctic wasteland. During the long, cold winters, they receive no light at all.

Diatoms are at the beginning of an extensive food chain that includes many of the world's creatures, including humans. *Zooplankton* are microscopic animals that feed on the diatoms. Small crustaceans such as krill feed on zooplankton. These animals, in turn, feed many of the whales and even humans. Without diatoms, the base of this particular food chain would collapse.

Considered alone, the great variety, beauty and creativity that can be seen in these creatures attest to an overall plan in the created order. However, the additional complex interrelationships, such as this food chain, bear additional witness of the great intelligence that oversees all the creation.

The "book of nature" can only hint at what we can learn about its Creator in the pages of His revealed Word, the Bible.

Prayer: Dear Father, though the creation is glorious, its witness to You is vague compared to Your revelation of Yourself in the Bible. Through my study of Your Word and my knowledge of Your creation, help me to be a bolder witness of You, especially to those who are led to believe in a vague "god of nature." In Jesus' Name. Amen.

REF.: Browne, Malcolm W. 1984. Antarctica: life's tenuous toehold. *Discover*, Mar. p. 48.

Those Astonishing Bee Engineers

Job 11:7-9

"Canst thou by searching find out God? canst thou find out the Almighty unto perfection? It is as high as heaven; what canst thou do? deeper than hell; what canst thou know? The measure thereof is longer than the earth, and broader than the sea."

The amazing structure of the honeycomb has fascinated scientists for thousands of years. In the third century, the astronomer and geometer Pappus of Alexandria became the first to offer an explanation for why the honeycomb has a hexagonal shape.

Pappus explained that only three shapes could serve as candidates for a honeycomb cell—the triangle, the square, and the hexagon. Any other shape would leave wasteful open spaces between the cells. Pappus noted that the hexagon holds more honey in the same space than either a square or a triangle. It also takes less wax to build and the shared sides of the hexagonal cells reduce wax usage even further.

It was not until the development of modern calculus that scientists could fully appreciate the shape of the caps at the end of the honeycomb cells. Each cell is capped with a pyramid composed of three rhombuses. Complex mathematics shows that this shape, too, requires the least amount of wax for construction. It also allows honeycomb cells to be butted up against each other without wasting space.

Modern scientists who accept evolution talk about the design of the honeycomb as a great accomplishment by the bees. However, the more sensible conclusion is obvious. The ten-sided prism of the honeycomb cap is magnificent testimony to the mathematical wisdom of the Creator Himself!

Prayer: Only You, Lord, could have invented mathematics and then taught it to the bees! Your wide generosity to all creatures enriches and blesses our lives even more. Amen.

REF.: The mathematics of the honeycomb. *Science Digest*, June, 1985. p. 74.

Your Busy Liver

Psalm 139:13

"For thou hast possessed my reins: thou hast covered me in my mother's womb."

An adult human liver is about the size of a small football. It weighs about three pounds, making it the body's largest internal organ. Tucked neatly beneath the ribs, your liver performs more than *500 different* tasks. It is a vital link between your heart, lungs and digestive system.

Inside the liver is a bewildering array of microscopic veins in which each drop of blood is processed. Here, blood conditions are constantly monitored to make sure that its chemistry meets strict standards. If more of some substances are needed in the blood, they are supplied. Useless chemicals are broken down into useful chemicals. Your liver makes proteins, and corrects blood clotting factors. Your liver also sees to it that hormone balances are maintained and that poisons are neutralized. If substances are needed to fight an infection, your liver makes them and adds them to the bloodstream. Your liver also stores vitamins and minerals and prepares itself to provide your body with quick energy when you need it. In addition, the liver makes bile, which is essential for digestion.

Structures like the liver have caused many evolutionists to abandon the idea that life is a result of millions of years of accidents. The liver is just too well-designed and integrated into the body to have been produced by purposelessness and mindlessness. As Scripture says, truly God *has* formed our inward parts!

Prayer: I thank You, dear Father, that I am so fearfully and wonderfully made. Help me to take good care of the wondrous body You have given me and grant me good health. In Jesus' Name. Amen.

REF.: The liver: the body's refinery. *Discover*, Apr. 1984. p. 80.

Who Said the Earth Was Flat?

Isaiah 40:22

"It is he that sitteth upon the circle of the earth, and the inhabitants thereof are as grasshoppers; that stretcheth out the heavens as a curtain, and spreadeth them out as a tent to dwell in:"

Many of us have heard the old legend about why Christopher Columbus was told he could not sail too far west. According to the legend, Columbus was repeatedly told that he could not sail west to India because the earth was flat. If he sailed too far, he would fall off the edge. That story was a piece of fiction cooked up by Washington Irving in the 1830s.

Only one writer can be found in Western history who believed the earth was flat—that writer was Lactantius in the third century AD. All the others, for example Pythagoras, still noted today for his mathematical proofs, simply assumed the earth was spherical. Two hundred years later, the Greek mathematician Eratosthenes, also assuming that the earth was round, calculated that the earth was 25,000 miles in circumference. He was off by only 200 miles. A century later, using Eratosthenes' figure for the earth's diameter, Hipparchus calculated that the moon was 240,000 miles away from the earth. He was off by only 100 miles—less than half a percent!

Nor does the Bible teach that the earth is flat. Isaiah 40:22 tells us how God looks down on the inhabitants of a round earth. Job 26:7 talks about the earth floating in space, hung on nothing.

Today we have exposed *two* myths. One is the myth that most people used to think the earth was flat. The other myth is that the Bible is not accurate when it describes the same creation that modern science talks about.

Prayer: I thank You, Lord, that Your Word is trustworthy. Human words are often in error or deliberate deceptions. Help me to seek a higher standard of truth, honesty and godliness in all my words, so that I may be more like You. Amen.

REF.: Lemonick, Michael D. 1985. Measuring the earth. *Science Digest*, July. p. 72.

The Marvelous Bat

Romans 11:33

"O the depth of the riches both of the wisdom and knowledge of God! how unsearchable are his judgments, and his ways past finding out!"

Bats are among the most underappreciated creatures in the entire world. Yet they are most worthy of our appreciation. Nearly all bats eat only insects or fruit. Only one tropical species feeds on blood—usually the blood of cattle.

There are thousands of bat species. Incredibly, bats make up one-fourth of *all* mammal species on earth. One-half of all mammal species found in the tropics are bats.

The smallest bat in the world is the Kitti's hog-nosed bat, which is no larger than a bumblebee and weighs less than a penny. The largest bats are fruit bats. These completely harmless bats may have a wingspan of several feet. Most bats live in colonies, and some colonies have up to 20 million members.

Bats are beneficial to both humans and nature. Many of the bats found in North America, as well as other parts of the world, are responsible for consuming thousands of tons of insects every night. Other bats pollinate flowers. Bats are also an essential link in the ecology that makes the rain forest possible.

The earliest bats found in the fossil record are like bats today. They are fully formed with fully developed wings for flight. This means that there is not one scrap of fossil evidence to support the claim that bats evolved from some other creature. We can conclude that the scientific evidence says that bats were created pretty much in their present form by God.

Prayer: Father, Your wisdom is its own witness. Help me to appreciate Your wisdom more as I seek it in Your Word. When I cannot understand what You are doing in my life, strengthen my faith and assure me of Your love. In Jesus' Name. Amen.

REF.: Silent screamers. *Science Digest*, May 1984. p. 57.

An Intelligent Plant

Matthew 6:27

"Which of you by taking thought can add one cubit unto his stature?"

Scientists have known for some time that hummingbirds prefer bright-colored flowers. On the other hand, all of us know that white is most easily seen at night. That means that moths that pollinate flowers will more easily find white flowers. The question is, what is a flower to do when it is pollinated by hummingbirds during part of its flowering season *and* by night-flying moths later in its flowering season?

That's the problem faced by a flowering plant called the gilia. Hummingbirds arrive to pollinate the flowers in mid-July. However, they leave in late August as the weather cools. Also in late August, the night-flying hawk moth arrives to begin pollinating the flower. This arrangement continues until late September. To attract both hummingbirds and moths during its long flowering season, the gilia provides bright red flowers, favored by the hummingbirds, until late August. Then it offers white flowers to attract the hawk moths.

Evolutionary scientists call this an example of evolutionary strategy. Think about that for a moment. How would a flower know what color it is? How would it know which colors are preferred by hummingbirds and which are preferred by moths? A flower could no more decide to change its color than you or I can decide to grow an inch taller.

Both logic and science say that the gilia's special features are intelligently designed by a truly gifted engineer. Why not, then, recognize the work of our all-wise Creator for what it is!

Prayer: I confess, dear heavenly Father, that I worry too much. I know that You are in charge of the universe and I am not. Forgive my worry for the sake of Your Son, my Lord and Savior, Jesus Christ, and strengthen my trust in You. Amen.

REF.: Plants change colors to lure pollinators. *Science Digest*, May 1985. p. 19.

Catastrophe!

Genesis 7:11

"In the six hundredth year of Noah's life, in the second month, the seventeenth day of the month, the same day were all the fountains of the great deep broken up, and the windows of heaven were opened."

The year of 1816 was called the "year without a summer." Weather records and newspaper clippings offer us a record of the unusual weather that struck all over the earth in 1816.

A Vermont newspaper reported that on June 6, it began to snow and hail at 10:00 in the morning and continued until evening. In mid-July, the corn crop in Maine was almost completely destroyed by frost. In late August, the remaining corn crop, as far south as Massachusetts, was frozen out along with other crops.

This bizarre story actually begins a year before, in 1816, in Indonesia. That's when Mt. Tomboro began to erupt with such violence that some of the explosions could be heard 1,000 miles away. In the end, Tomboro had blown over 25 cubic miles of debris into the atmosphere. This dust blocked enough of the sun's light to virtually cancel the following summer all over the earth.

However, the lesson for us is that huge, earth-changing catastrophes are a normal part of our planet's history. It might take normal processes 500 years to create a layer of sediment six inches thick. However, a flood, volcanic eruption or hurricane can create the same layer in only a few hours. Just think how far off evolutionary dating is when this happens. We know from experience that things don't just continue for thousands of years without some catastrophe taking place.

Prayer: Dear Father in heaven, the violence and power of an earthquake or volcanic eruption reminds us how helpless we truly are. Increase my conviction that my life always rests in Your loving hands so that the violence of nature does not have to remind me. In Jesus' Name. Amen.

REF.: Stommel, Henry, and Elizabeth Stommel. 1980. The year without a summer. *Scientific American*, Dec. p. 176.

How to Engineer a Faster Shark

Job 9:10

"Which doeth great things past finding out, yea, and wonders without number."

Scientists studying fluid dynamics are trying to learn how to create a ship or submarine hull that has the least drag in the water. Reducing drag not only saves fuel, it allows a ship or submarine to move more rapidly through the water.

As a hull moves through the water, eddies are created in the layer of water next to the hull. When they first form, these eddies flow smoothly, adding little drag. As they continue to form, they coalesce into what scientists call a "burst." A "burst" is an area where the eddies gang up to draw energy away from the speeding hull into the surrounding water.

If scientists could find a method of reducing drag on hulls, they could save fuel, billions of dollars, and time. On the sea, time often equals lives. In one breakthrough, scientists learned that tiny ribs or grooves on the surface of an object—called "riblets"—help prevent the formation of bursts. The "riblet" design can reduce drag up to 10%.

Most interesting is that the skin of sharks is also designed on the "riblet" principle. This makes the shark even more efficient than our fastest ships.

The "riblet" design is not hard to explain for those who accept that the shark was made by a Creator. The Creator of water as well as sharks would have a complete understanding of these principles that are only now being learned by humans.

Prayer: Dear Lord, I thank You that we are able to learn and benefit from the principles You have built into the creation. I ask that You would move more Christian young people to go into science and that You would guard and protect their faith from the forces of unbelief. Amen.

REF.: Sharks and speed. *Discover*, Apr. 1984. p. 70.

Humans Are More than Matter

Zechariah 12:1

"The burden of the word of the LORD for Israel, saith the LORD, which stretcheth forth the heavens, and layeth the foundation of the earth, and formeth the spirit of man within him."

Evolutionary scientists studying the workings of the human brain are trying to find out what the human "mind" is. The problem is that everyone, including scientists, knows that human beings are conscious people with minds. Yet strict, materialistic science says that there is no scientific evidence for the mind.

Research with people who had brain injuries indicates that the brain and the mind are not the same things. The reality of the human mind seems to have no material basis. This presents a difficult problem for evolutionary scientists who have assumed that evolution, with its materialistic view of reality, is correct. According to the evolutionary view of the world, if something is not measurable by science, that thing is not real. Evolution denies that humans have any nonmaterial part.

Some evolutionists have tried to resolve this problem by saying that human beings are not really conscious. They say that we don't really have minds. We only *think* we do. It's a trick played on us by our nervous systems. Entire books have been devoted to this subject. However, this approach creates a new problem. If we really don't have minds, we cannot trust the truth of *anything* we might think. We can't even trust the thought that we don't have minds.

The debate among these strict materialist scientists actually amounts to an admission that there is a greater, nonmaterial reality that cannot be measured by science. Yes, the existence of the immaterial, but undeniably real, human mind (should I say "soul"?) also testifies to the Creator!

Prayer: Dear Father, I thank You that You have made it so evident that human beings are more than simply material organisms. Strengthen Your people on earth so that they can witness to Your truth, which truly feeds the soul. In Jesus' Name. Amen.

A Plague of Biblical Proportions

Exodus 10:4

"Else, if thou refuse to let my people go, behold, to morrow I will bring the locusts into thy coast."

When Pharaoh refused to let the Israelites leave Egypt, the eighth plague the Lord brought upon Egypt was locusts. Moses warned stubborn Pharaoh that the locusts would be so thick that no one would be able to see the ground (Ex. 10:4-6). Scripture also makes clear that there never had been, nor would there ever be again, such a great infestation.

We gain an appreciation for what the Egyptians had to endure because of the hard-heartedness of their leader when we look at modern plagues of locusts. A locust weighs only two grams, but it needs to eat its own weight in food every day. That doesn't seem like much. However, a swarm of locusts can cover an area of 400 square miles with a density of 100 million to 200 million insects per square mile! A swarm this size can eat 80,000 tons of food per day. That's enough to feed 40,000 people for a year.

Locusts can descend on an area without warning almost as fast as they did in Egypt. Desert locusts have been known to fly nonstop for up to 1,600 miles at a time.

This knowledge doesn't take away from the miraculous acts of God recorded in the Bible. Knowing a little about the power of a normal locust infestation helps us appreciate the plague of locusts on Egypt. As the Bible says, this locust infestation was unlike any other in history!

Prayer: Dear Lord, I ask that You would protect crops, especially in areas where food is scarce. More importantly, I ask that You would control events so that Your people could bring the witness of the Gospel to those who are spiritually starving. Amen.

REF: London, Jane. 1979. *Science Digest*, June. p. 25.

The Bowerbirds of New Guinea

Ecclesiastes 3:11

"He hath made every thing beautiful in his time: also he hath set the world in their heart, so that no man can find out the work that God maketh from the beginning to the end."

The bowerbirds of New Guinea and Australia are among the most unusual birds in the world. The group of species known collectively as bowerbirds are named from the bowers, or small huts, that the males build in the hope of wooing a mate.

These bowers are constructed of sticks, stood umbrella fashion around a center support pole. They are usually built on green moss right on the ground. Most bowers are around two and one-half feet tall and six feet in diameter. They are so well built by the birds that they are waterproof.

Once construction is complete, the bowerbird sets out to collect colorful decorations. He uses these to decorate the inside of his home and his front yard. Decorations include fruits, leaves, flowers, butterfly wings, and even dead beetles. The more colorful the bauble, the better. In decorating his front yard or inside his home, same-colored items are always stacked together, and decorations are moved and replaced daily. This is done in the hope that some nice lady bowerbird will come by and be attracted by his sense of interior and exterior design.

It's interesting that, contrary to what we often hear, there *is* a common sense among creatures of what is beautiful. The bowerbird's colorful arrangements are attractive to bowerbirds as well as to humans. This suggests that we share a common sense of beauty with other creatures because our Creator has an appreciation of beauty.

Prayer: I thank You, Lord, for all the beauty that surrounds us. Your life, death and resurrection has made my life beautiful to the Father through the forgiveness of my sins. Help me to live this new life You have given me as a beautiful offering to Him. Amen.

REF.: Diamond, Jared. 1984. The bower builders. *Discover*, June. p. 52.

A Most Colorful Artist

Matthew 10:29

"Are not two sparrows sold for a farthing? and one of them shall not fall on the ground without your Father."

The female bowerbirds of New Guinea and Australia select their mates based on the bowers, or umbrella-shaped huts, built and decorated by the males.

Females appear to judge a male on both the construction and design of his bower as well as on his interior and exterior decorating. Male bowerbirds collect flowers, snail shells, leaves and even butterfly wings to create their elaborate color arrangements in and around their bowers. Scientists studying bowerbirds in New Guinea found that even their own brightly colored items could end up in some bird's decorating scheme if they are left unguarded.

Bowerbirds are also very individualistic in their decorating tastes. Scientists found one bowerbird who never used anything yellow in his schemes. This bird would even steal decorations from other bowerbirds. You could always tell when this little thief struck a neighboring bower, because nothing but yellow decorations would remain. Some bowerbirds specialize in butterflies, and one specialized in nothing but purple flowers.

Our Lord Jesus Christ told us that not one sparrow falls to the ground without the knowledge of the Heavenly Father. He said this to point out that all creatures are individuals. Each is sought and loved as an individual by his or her Creator. Like us, individual bowerbirds express their individuality in their varying tastes. However, what about the second part of Christ's lesson? Do *you* know the love of your Creator for you through Jesus Christ?

Prayer: I thank You, Father, that You have made me an individual. I don't understand how You see and treat me as an individual with billions of other people in the world. Yet, I believe Your Word. Help me when I have difficulty believing in Your individual love and care for me. In Jesus' Name. Amen.

REF.: Diamond, Jared. 1984. The bower builders. *Discover*, June. p. 52.

Tower-Building Birds

Psalm 27:4

"One thing have I desired of the LORD, that will I seek after; that I may dwell in the house of the LORD all the days of my life, to behold the beauty of the LORD, and to inquire in his temple."

Most male bowerbirds build a hut about two and one-half feet high and about six feet in diameter.

However, scientists have discovered two different groups of *tower*-building bowerbirds. These unique bowerbirds are found in the Kumawa mountains in New Guinea. Each bird builds up to five towers, each up to nine feet high, hoping to attract a female. Tower-builders don't simply weave sticks together, as do other bowerbirds. Rather, as they weave the sticks together, they also glue them, possibly using their saliva. These substantial constructions can weigh up to 70 pounds.

The Kumawa tower-builders decorate their yards and the insides of their towers with variously colored items. However, tower-builders prefer different decorations than most bowerbirds. One of the favorite decorations is the pandanus leaf, which can be from three to six feet long. These leaves are leaned up against the towers. Other decorations may include 1,000 or more snail shells, acorns, black stones and white stones.

Evolutionists admit that they are mystified about how to explain the behavior of bowerbirds. Evolution can't explain this strange behavior, nor can it explain the bowerbirds' taste for beauty and color. However, if we recognize bowerbirds as the work of the Creator God, who also made sunrises and sunsets, bowerbirds remain a wonder, but they are no mystery.

Prayer: I thank You, Lord, that even after sin has ruined Your creation, so much beauty still shows through. Grant me a greater appreciation for the spiritual beauty of Your truth and gifts of grace. Amen.

REF.: Diamond, Jared. 1984. The bower builders. *Discover*, June. p. 52.

Does Biology Rule Out Christmas?

Matthew 1:23

"Behold, a virgin shall be with child, and shall bring forth a son, and they shall call his name Emmanuel, which being interpreted is, God with us."

Christians around the world believe that Jesus Christ was born of the Virgin Mary. To those who don't believe in a Creator, the virgin birth of God in human flesh is beyond human comprehension. However, why should science deny what it can never understand?

Life is filled with intervention by intelligent forces that we cannot understand. Television allows us to watch what is happening this instant 10,000 miles away. Few, if any of us, understand every one of the intelligently directed efforts that make television possible. Who could master and control all the details that bring together satellite, television, sound and electrical control technologies to produce sights and sounds from anywhere in the world? However, our inability to understand television does not lead us to deny its existence.

Then can we say that God cannot take the form of His creatures simply because He loved us? Can we say that it is impossible for a virgin to give birth to God in the flesh? Can science rule out His suffering and death for the payment of our sin? Can science presume to tell us His grave is not empty?

Even if God would have visibly shown skeptical scientists every step, our measures are too narrow to span the width, too small to plumb the depth of His purposes. And for that we thank Him. We thank Him for intervening in our history with the birth of His Son. Truly the depths of God's acts are only comprehended in heavenly light!

Prayer: Father, I am awed and filled with thanksgiving at the mystery of Your love. Help me to be a bold and fearless witness of the forgiveness of sins in Jesus Christ to those around me. In His Name. Amen.

The Creating Word Made Flesh

Ephesians 1:3-4

"Blessed be the God and Father of our Lord Jesus Christ, who hath blessed us with all spiritual blessings in heavenly places in Christ: According as he hath chosen us in him before the foundation of the world, that we should be holy and without blame before him in love:"

The story of Christmas begins with the beginning of the work of the Son of God. And His first work is recorded in Genesis chapter 1. The story of His love for us that led Him to give up the glory, beauty and perfection of heaven to come to our dirty, dusty, pain-filled world begins not in Bethlehem, or in the Old Testament prophets. We must go to the beginning of His-story lest His birth become simply the beginning of another story of a tragic and misunderstood life.

The cross of Golgotha stands firmly planted in Genesis chapter 1. Even before the creation, Christ's pain and suffering on that coarse structure was anticipated. Ephesians 1:4 reveals that even before the foundation of the world God's plan of salvation was already in place.

Even the coarse Roman soldiers knew that His robe could not be cut into pieces without destroying it. Likewise, the Bible is one complete piece—woven of one thread. Cutting out anything leaves the whole thing to unravel. To remove the beginning of the story of the Word and His work in Genesis is to remove the foundation on which His work of salvation stands.

Yes, the Christmas story is far more than the story of the birth of an important baby. It is the center of the story of creation itself!

Prayer: I thank You, Lord, that You made me. And I thank You that when I was lost in sin and Your enemy, You sought me out with a plan devised before the world was formed. Help me to love others as You have loved me. Amen.

REF.: Bartz, Paul A. 1984. The creating world made flesh. *Bible-Science Newsletter*, Dec. p. 1.

The Christmas Star

Matthew 2:9

"When they had heard the king, they departed; and, lo, the star, which they saw in the east, went before them, till it came and stood over where the young child was."

Around Christmastime it is common to hear some authority trying to explain the star that the wise men followed to find the infant Jesus. It is commonly explained as a comet, or the conjunction of two or more planets or stars. The question is, are these and similar theories possible?

About all you need to evaluate these theories are a Bible and a map of the Holy Land. Matthew 2:1 tells us that the wise men came from east of the Holy Land. Verses 2 and 9 further tell us that these men had seen the star indicating His birth in the *eastern* sky. They had *not* followed the star to Jerusalem, as often depicted in Christmas cards. When they arrived in Jerusalem, we learn from Matthew 2:9 that the star they had seen in the eastern sky *moved* to lead them to the Christ Child.

By this time, Mary, Joseph, and the Child may already have returned to Nazareth, or they may still have been in Bethlehem. No matter which was the case, the movement of the star was now from the eastern sky to either the northern or the southern sky. That is very un-starlike. In fact, the movement of the star was even more un-starlike in that it directed them to a *specific house*! No comet, conjunction or other natural object of the night sky will do that.

These wise men from the East were accustomed to the sights of the night sky. They could easily identify the unusual. This star was clearly created by God to announce the birth of the Savior to the world He came to save!

Prayer: Dear Father in heaven, You create and control powers I could never comprehend. I thank You that You led the wise men to the infant Jesus, clearly showing that His birth was for all of us on earth. Thank You for bringing me into His kingdom. Amen.

Nobody's Perfect

2 Peter 3:17-18

"Ye therefore, beloved, seeing ye know these things before, beware lest ye also, being led away with the error of the wicked, fall from your own stedfastness. But grow in grace, and knowledge of our Lord and Saviour Jesus Christ. To him be glory both now and for ever. Amen."

Scientists, like everyone else, can make errors. Science has its own "hall of shame." Because what scientists say is often printed in textbooks and repeated as fact, errors in science tend to have a bigger influence on people's thinking.

Some errors in science are humorous. For example, it was a scientific error more than 80 years ago that led generations of mothers to think that spinach was full of iron and good for their children. The error arose when a turn-of-the-century scientific paper misplaced the decimal point in its report on the amount of iron in spinach. No one ever questioned the error. After all, if most children don't like spinach, it must be good for them. In fact, not only does spinach have very little iron, but it can be harmful to children, because it tends to remove precious bone-building calcium. But children—don't rejoice just yet. Raw oysters are very rich in iron.

Other errors in science are not so amusing. The idea that a developing embryo goes through supposed earlier evolutionary stages has been used to promote both evolution and abortion. Yet science rejected this idea generations ago. The claim that *Archaeopteryx* was an evolutionary transitional form between reptiles and birds is another scientific error that has been used to promote evolution and call the Bible's history into question.

Yes, everyone makes mistakes. However, evolutionists have committed more than their fair share in attacking the truth of the Bible.

Prayer: I recognize, Lord, that I am not perfect. However, I know that You lived a perfect life on my behalf and then suffered my punishment for sin on the cross. Help me not to return to justifying myself, but help me seek justification in You. Amen.

REF.: I say it's spinach. *Science Digest*, Feb. 1979. p. 96.

God's Answer to Indoor Pollution

Psalm 107:8

"Oh that men would praise the LORD for his goodness, and for his wonderful works to the children of men!"

News reports about indoor pollution are making more people concerned about what chemicals may be in the air in their own homes. Reports say many homes have high levels of formaldehyde from particle board. Benzene from household cleaners can cause dizziness, headaches and even nausea.

Imagine, then, the problem faced by the National Aeronautics and Space Administration when tests on Skylab showed that more than 100 trace amounts of chemicals polluted the air of that space station.

You might guess that NASA began looking for the answer to the problem in high technology filters. However, technology did not provide the best solution to the problem. NASA found the answers to indoor pollution working jointly with the Associated Landscaping Contractors of America. They learned that many easy-to-care-for houseplants do an incredible job of clearing the air of even the most exotic pollutants. In one test, a spider plant cleared a test chamber of 85% of all pollutants in only 24 hours! NASA has also learned that the Chinese evergreen, the peach lily, and golden pothos are also very good at clearing indoor pollutants.

This means that your home can have some of the same inexpensive, efficient air filters used by NASA. It is obvious that God, knowing beforehand what our needs would be, created these easy-to-grow plants with the unique ability to clear the air and help keep us healthy!

Prayer: Father, You know our needs even before we do. Help me never to tire of seeking help from You for both my material and spiritual needs. In Jesus' Name. Amen.

Letting God Create Your Day, Volume 2 Index